Lean Maintenance

Lean Maintenance

Ricky Smith
Bruce Hawkins

ELSEVIER
BUTTERWORTH
HEINEMANN

AMSTERDAM • BOSTON • HEIDELBERG • LONDON
NEW YORK • OXFORD • PARIS • SAN DIEGO
SAN FRANCISCO • SINGAPORE • SYDNEY • TOKYO

Elsevier Butterworth–Heinemann
200 Wheeler Road, Burlington, MA 01803, USA
Linacre House, Jordan Hill, Oxford OX2 8DP, UK

 Recognizing the importance of preserving what has been written, Elsevier prints its
books on acid-free paper whenever possible.

Library of Congress Cataloging-in-Publication Data
Smith, Ricky.
    Lean maintenance : reduce costs, improve quality, and increase market share /
Ricky Smith and Bruce Hawkins.
        p.   cm.—(Life cycle engineering)
    Includes index.
    ISBN 0-7506-7779-1
    1. Production management.   2. Manufacturing processes.   3. Just-in-time systems.
I. Hawkins, Bruce. II. Title.   III. Series.
    TS155.S635 2004
    658.5—dc22   2003026393

British Library Cataloguing-in-Publication Data
A catalogue record for this book is available from the British Library.

ISBN: 0-7506-7779-1

For information on all Butterworth–Heinemann publications
visit our Web site at www.bh.com

04 05 06 07 08 09 9 8 7 6 5 4 3 2 1

Printed in the United States of America

Dedication

Contributing Authors to the Lean Maintenance book were Randy Heisler, Dan Dewald, and Erich Scheller.

Thanks to Joe Dvorak and Kevin Campbell, with Northop Grumman Newport News Shipyard, for providing us with the stimulation to write this book.

Thanks to John Day, formerly with Alumax, for giving us the vision of true maintenance.

Very special thanks to Bill Klein for all of his contributions to this book. Without Bill this book would not be possible.

We also want to offer thanks to Jim Fei, Chairman and CEO of Life Cycle Engineering, Inc. Without Jim's understanding and commitment to the engineering and maintenance community, this could not have happened.

Contents

Preface

Lean Maintenance is a relatively new term, coined in the last decade of the twentieth century, but the principles are well established in Total Productive Maintenance (TPM). Lean Maintenance—taking its lead from Lean Manufacturing—applies some new techniques to TPM concepts to render a more structured implementation path. Tracing its roots back to Henry Ford with modern refinements born in Japanese manufacturing, specifically the Toyota Production System (TPS), Lean seeks to eliminate all forms of waste in the manufacturing process—including waste in the maintenance operation. While the first chapter of this Lean Maintenance Handbook seems to dwell on Lean Manufacturing and does not address maintenance, there is a purpose for that. All Lean thinking—the premise of Lean Manufacturing and Lean Maintenance—is originally based on manufacturing processes. Some believed that everything else would just naturally evolve, or fall into line, from those roots. Time, however, has unmasked the difficulties of instituting "Lean" in production support operations, those areas adjacent to the manufacturing production process, such as maintenance, without the presence of some prerequisite conditions. That topic is the subject of the remainder of this book after initially establishing some common ground.

To reduce costs and improve production, most large manufacturing and process companies that have embraced the Lean Enterprise concept have taken an approach of building all of the systems and infrastructure throughout the organization. The result of this traditional approach has been erratic implementation efforts that often stall-out, or are terminated, before the benefits come. Plants can accelerate their improvements with much lower risk through the elimination of the defects that create work and impede production efficiency. Optimizing the maintenance function first will both increase maintenance time available to do further improvements and will reduce the defects that cause production downtime. Thus cost reduction and improved production are immediate results from establishing Lean maintenance operations as the first step in the overall Lean Enterprise transformation.

Lean Maintenance is intended to be a stand-alone teaching text that provides the reader with all the terminology (defined), all of the Lean Implementation Processes—including techniques for getting the most from the application of each process—and all of the planning and sequencing requirements for proceeding with the Lean Maintenance Transformation journey—including methodologies and background information. At the same time, or rather after it has served its purpose as a teaching text, *Lean Maintenance* is intended to be a quick-reference volume to keep with you during your actual journey through the Lean Transformation. We have tried, through the extensive use of charts, tables, and checklists, to make any single piece of information, as well as the sum of all of the information, simple to locate and effortless to understand.

1

Common Ground

1.1 THE HISTORY AND EVOLUTION OF LEAN

1.1.1 Manufacturing Evolves

Manufacturing is the conversion of raw materials, by hand or machine, into goods. Craftsmen were the earliest manufacturers. Highly skilled craftsmen, even today, spend several years in apprenticeship learning their craft. They often made their own tools and sold their own finished goods. Craftsmen obtained and prepared their basic raw materials, moved the product through each of the stages of manufacture and ended with the finished product. Because of the time involved producing the finished goods, craftsmen-manufactured products were, and are, costly.

The Industrial Revolution brought about the division of labor, a specialization of focused and more narrow skills applied to a single stage of the manufacturing process. Generally acknowledged as beginning around 1733, with John Kay's invention of the "flying shuttle" for the textile industry, the Industrial Revolution brought about tremendous changes in society with the creation of the working class. The working class earned, and spent, an income on a continual, year-round basis, a significant departure from the previously agrarian society.

Followed by several important inventions between 1733 and 1765, when the steam engine was perfected by James Watt and ultimately applied to the cotton milling industry in 1785, the replacement of human, animal, or water power by machine assisted motive power solidified the concept of mass production. The introduction of machines to manufacturing soon brought about the manufacture of goods with interchangeable parts.

By the mid-to-late 1800s, the concepts of division of labor, machine-assisted manufacture, and assembly of standardized parts to produce finished goods, were firmly established in Europe and the United States. Large factories were appearing in all the urban areas because of the need for large numbers to make up the labor force. Mass production of various goods created the beginnings of society's first "middle class." In these early stages, the methods used to organize labor and control the flow of work were less than scientific, based primarily on precedent and historical usage rather than on efficiency.

In 1881, Frederick W. Taylor began delving into the organization of manufacturing operations at the Midvale Steel Company, and Industrial Engineering was born. His refinement of methods and tools used in the various stages of steel manufacture permitted workers to produce significantly more with less effort. Shortly after the turn of the century, further refinements were made by Frank B. Gilbreth and following his death, continued by his wife, Lillian M. Gilbreth (both of whom gained a measure of notoriety in the biographical novel and motion picture *Cheaper by the Dozen*) through time-motion studies, which brought about a quantitative approach to the design of contemporary manufacturing systems and processes.

When Henry Ford was born in 1863, Abraham Lincoln was president. By 1896, Ford had completed building his first horseless carriage, which he sold to finance work on a second vehicle, and a third and so on. In 1903 the Ford Motor Company was incorporated with $28,000 in cash invested by ordinary citizens. In 1908 Ford announced, "I will build a motor car for the great multitude" as he unveiled the Model T. During the nineteen years of the Model T's existence, Ford sold nearly seventeen million of the cars—a production total amounting to half the automobile production of the entire world.

This remarkable accomplishment was brought about by the most advanced manufacturing technology yet conceived—the assembly line. The assembly line employed the precise timing of a constantly moving conveyance of parts, subassemblies and assemblies, creating a completed chassis every 93 minutes.

This level of assembly line efficiency didn't happen overnight. It took more than five years of fine-tuning the various operations, eliminating the wasted time in each of them to reduce the initial assembly time of 728 minutes to the 93-minute output rate achieved in 1913. Manufacturing technology had just had its first encounter with lean thinking.

1.1.2 The Influence of Henry Ford

During its first five years, the Ford Motor Company produced eight different models, and by 1908 its output was 100 cars a day. The stockholders were ecstatic, but Henry Ford was not satisfied, believing he should be turning out 1,000 a day. The stockholders seriously considered court action to stop him from using profits to expand. In 1909, Ford, who owned 58% of the stock, announced that he was only going to make one car in the future, the Model T. The only thing the minority stockholders could do to protect their dividends from his all-consuming imagination was to take him to court, which is precisely what Horace and John Dodge did in 1916 (see Figure 1-1 below).

The Dodge brothers sued Ford for what they claimed was his reckless expansion and for reducing prices of the company's product, thereby diverting money from stockholders' dividends. The court hearings gave Ford a chance to expound his ideas about business. In December 1917 the court ruled in favor of the Dodges. The court said that, while Ford's sentiments about his employees and customers were nice, businesses were operated for the profit of its stockholders.

In March 1919, Ford announced a plan to organize a new company to build cars even cheaper than the Model T. Ford said that if he was not master of his own company, he would start another. The stockholders knew that without Henry Ford the Ford Motor Company would fail within the year. The ruse worked; by July 1919 Ford had bought out all seven minority stockholders. Ford Motor Company was reorganized under a Delaware charter in 1920 with all shares held by Henry Ford and other family members. Never had one man so completely controlled a business enterprise so gigantic.

1.1.2.1 Waste—The Nemesis of Henry Ford

Ford planned and built a huge new plant at River Rouge, Michigan. What Ford dreamed of was not merely increased capacity but complete self-sufficiency. World War I, with its shortages and price increases, had convinced him of the need to control raw materials. Slow-moving and unresponsive suppliers convinced him that he should make his own parts. Wheels, tires, upholstery, and various accessories were purchased from other companies around Detroit. As Ford production increased, these smaller operations had to speed their output; most of them had to install their own assembly lines. It became impossible to coordinate production and shipment so that each product would arrive at the right place and at

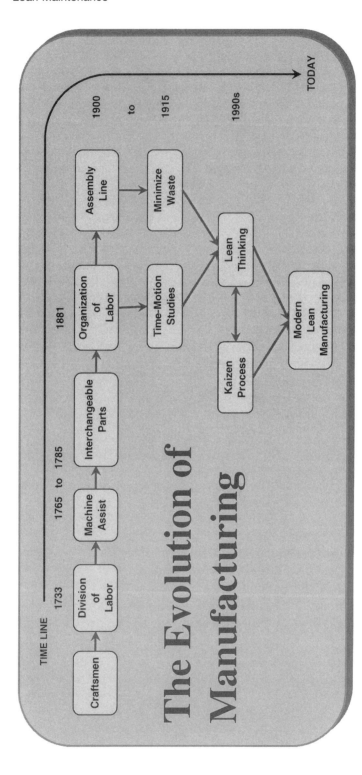

Figure 1-1 The Evolution of Manufacturing

the right time. At first he tried accumulating large inventories to prevent delays or stoppages of the assembly line, but he soon realized that stockpiling wasted capital. Instead, he took up the idea of extending movement to inventories as well as to production.

Ford believed that his costs in manufacturing began the moment the raw material was separated from the earth and continued until the finished product was delivered to the consumer. The plant he built in River Rouge embodied his idea of an integrated operation encompassing materials supply, component production, assembly and transportation. To complete the vertical integration of his empire, he purchased a railroad, acquired control of 16 coal mines and about 700,000 acres of timberland, built a sawmill, acquired a fleet of Great Lakes freighters to bring ore from his Lake Superior mines and even bought a glassworks.

Ford's primary objective remained building automobiles as inexpensively as possible. While control of all the inputs to the automobile manufacturing process didn't guarantee low costs, it did guarantee that Ford could manage all those input processes himself. He was all but obsessed with the elimination of waste. Wasted money, wasted material, wasted motion and wasted time all drove up the cost of his automobile, the cost per unit. Ford strove to purge waste from all levels in his vertically tiered manufacturing operation.

The move from Highland Park to the completed River Rouge plant was accomplished in 1927. At 8 o'clock every morning, just enough ore for the day's production would arrive on a Ford freighter from Ford mines in Michigan and Minnesota and would be transferred by conveyor to the blast furnaces and transformed into steel with heat supplied by coal from Ford mines in Kentucky. It would continue on through the foundry molds and stamping mills and exactly 28 hours after arrival as ore, it would emerge as a finished automobile. Similar systems handled lumber for floorboards, rubber for tires and so on. At the height of its success the company's holdings stretched from the iron mines of northern Michigan to the jungles of Brazil, and it operated in 33 countries around the globe.

Other automobile manufacturers were learning from Henry Ford. Notably, General Motors and Chrysler had established very viable manufacturing operations and were encroaching on Ford's dominant market share. One of the very premises of Ford's early success was in part to blame for this. He was still making the Model T and very cheaply (the Model T cost $950 in 1908 and $290 in 1927), but the automobile buyer's tastes were becoming more diversified.

When Ford finally became convinced that the marketplace had changed and was demanding more than a purely utilitarian vehicle, he shut down his plants for five months to retool. In December 1927 he introduced the Model A. The new model enjoyed solid but not spectacular success. Ford's stubbornness had cost him his leadership position in the industry; the Model A was outsold by General Motors' Chevrolet and Chrysler's Plymouth and was discontinued in 1931. Despite the introduction of the Ford V-8 in 1932, by 1936 Ford Motor Company was third in sales in the U.S. auto industry. Many believed that Ford had completely lost sight of his early values and that his objective had slowly evolved into one of dominance.

Perhaps they were right but regardless, no one can deny the tremendous influence that Ford had wielded throughout the entire field of manufacturing in the early twentieth century. Later in the century that influence was to have a profound effect on the Japanese.

1.1.2.2 Ford's Influence on Japanese Manufacturing

In 1936 in Yokohama, Japan, the Ford Motor Company was building Model A cars and trucks with mixed models in a plant converted over from the Model T. Ford was the largest automobile manufacturer in Japan in 1936. That same year, Sakichi Toyoda, owner and founder of Japan's largest loom manufacturing operation, started up a Japanese automobile manufacturing operation. As managing director of the new operation, Sakichi's son, Kiichiro Toyoda, traveled to the Ford Motor Company in Detroit for a year of studying the American automotive industry. Kiichiro returned to Japan with a thorough knowledge of the Ford production system. He was determined to not only adapt the system to smaller production quantities, but also to improve on the basic practices. In addition to the smaller production quantities, Kiichiro's system provided for different processes in the assembly sequence of production. His system managed the logistics of materials input to coincide with production consumption. Kiichiro developed an entire network of suppliers capable of supplying component materials as needed. The system was referred to as Just-in-Time (JIT) within the Toyoda Group.

In 1950, when the Toyoda Group was forced by the Japanese government to reorganize, Eiji Toyoda was named the new managing director. Kiichiro resigned and his cousin Eiji took hold of the company's reins. Like Kiichiro, Eiji also went to the United States to study the American system of automobile manufacturing. Among the concepts that Eiji brought back to Japan was Ford's suggestion system. Not content to simply copy American practices, Eiji Toyoda instituted the first Kaizen (continuous improvement) process within the Toyoda Group based on the Ford Motor Company's suggestion system. In 1957, Eiji renamed the company Toyota and in the same year opened a U.S. sales operation.

Taiichi Ohno joined the Toyoda organization after graduating from Nogoya Technical High School in 1932. Early in his career, he expanded upon the JIT concepts developed by Kiichiro Toyoda to reduce waste, and started experimenting with and developing methodologies to produce needed components and subassemblies in a timely manner to support final assembly. During the chaos of World War II, the Toyoda Loom Works, where Ohno worked, was converted into a Motors Works and Taiichi Ohno made the transition to car and truck parts production. The war resulted in the leveling of all Toyoda Group Works production facilities, but under the management of Eiji Toyoda, the plants were gradually rebuilt and Taiichi Ohno played a major role in establishing the JIT principles and methodologies he had helped develop and further refine in the Loom manufacturing processes.

At the reconstructed Toyoda Group Automotive Operations, renamed as the Toyota Automobile Group, Taiichi Ohno managed the machining operations under severe conditions of material shortages as a result of the war. Gradually he developed improved methods of supporting the assembly operations. The systems that were developed (the Toyota Production System, or TPS), Ohno credited to two concepts brought back from the United States. The first concept was the assembly line production system. Ohno derived the system from Henry Ford's book *Today and Tomorrow,* first published in 1926. The second concept was the supermarket operating system in the United States, which Ohno observed during a visit in 1956. The supermarket concept provided the basis of a continuous supply of materials as the supermarket provides a continuous supply of goods to the consumer. This pull system replenished items as consumers purchased (or pulled) them from the supermarket shelves. Today Taiichi Ohno is widely acknowledged as the father of the Toyota Production System— Lean Manufacturing.

Additionally, significant influence in shaping the Toyota Production System was provided by Shigeo Shingo, a quality consultant hired by Toyota, who assisted in the implementation of quality initiatives; and W. Edwards Deming who brought Statistical Process Control to Japan. The Toyota Production System embodies all of the present day attributes of Lean Manufacturing.

In short, Lean practices are the practices of waste elimination and continuous improvement.

1.1.2 Japan's Refinement of Ford's Mass Production System

The principles and practices of Lean, although very basic and fundamental in concept, were developed over a 90-year period of time. While they have evolved by trial and error over many decades, and many prominent men have contributed to their refinement, the principles and practices are not easily implemented. Implementation requires a commitment and support by management, and active participation of all of the personnel within an organization to be successful.

Japan did not invent Lean practices with the Toyota Production System (TPS). They adapted what they had learned from American automobile manufacturers, primarily Henry Ford, and from other American industries. What they did was to apply their uncanny ability to focus intently and single-mindedly on root importances within a process while rejecting the unimportant aspects. After isolating the important they set about to not only improve, but to perfect those important aspects of American manufacturing concepts. These concepts included:

- Waste Elimination
- Standardized Work Practices
- Just-in-Time Systems
- Doing It Right the First Time (Quality Control)

Lean Manufacturing, as practiced within the TPS is a performance-based process used to increase their competitive advantage. The TPS employs continuous improvement processes to focus on the elimination of waste or non-value added steps in the manufacture and assembly of their vehicles. In perfecting the American manufacturing concepts cited above, the Japanese expanded and added a few, including:

- Integrated Supply Chain (from JIT)
- Enhanced Customer Value (from Quality Control)
- Value Creating Organization
- Committed Management
- Winning Employee Commitment/Empowering Employees
- Optimized Equipment Reliability
- Measurement (Lean Performance) Systems
- Plant-Wide Lines of Communication
- Making and Sustaining Cultural Change

Some of the more significant characteristics that they imbued their system with and the tools they developed, primarily within the Toyota Production System, in their pursuit of perfection are introduced briefly in the following sections of Chapter 1. Later chapters will address in detail the application of these, and other, characteristics, tools and methods as applied to Lean Maintenance practices.

Throughout this text you will be introduced to terms, many identified by their Japanese origin, which may be new and unknown to you. Please refer to the Glossary at the end of the handbook for complete definitions of these terms—as they are applied in this book.

1.1.2.1 The Kaizen Process

Literally, Kaizen means "continuous improvement." A Kaizen Event is the application of Kaizen (or Lean) tools to individual, small scale, one-week projects within the overall manufacturing operation. Each project is often referred to as a Kaizen Blitz Event. Kaizen Events are not applied in a single, plant-wide implementation of "Lean," but one event at a time. The Kaizen tools and methods, or processes, used in the execution of a Kaizen Event include the:

- 5-S Process
  1. Sort (remove unnecessary items)
  2. Straighten (organize)
  3. Scrub (clean everything)
  4. Standardize (standard routine to sort, straighten and scrub)
  5. Spread (expand the process to other areas)
- Identify and Eliminate the Seven Deadly Wastes
  1. Overproduction
  2. Waiting
  3. Transportation
  4. Processing
  5. Inventory
  6. Motion
  7. Defects
- Standardized Work Flow (TAKT [cycle] Time, work sequence and WIP [work in progress])
- Value Stream Mapping/Process Mapping (Use of symbols to draw a map of the steps in a process—Process Mapping)
- Kanban (Visual Cues or Signals)
- Jidoka (Perfection [Quality] at the Source—quality built in, not inspected in)
- Poka Yoke (Mistake or Error Proofing)
- Use of JIT and Pull (Supplying items JIT [Just-in-Time] and Pulling items only as you need them)

Not all of these terms (and perhaps none of them) may be familiar to you; however they will become integrated into your vocabulary once the decision to go "Lean" is made at your plant. Rather than define each one of them here, so that you can quickly forget them, each will be probed in depth as we describe Lean Maintenance and application of the Kaizen process, in particular to maintenance, in later chapters. One thing is important to remember before we proceed.

Lean thinking and Lean practices, in spite of a long list of complex and unfamiliar terms and names, are overwhelmingly just the application of common sense.

Kevin S. Smith, President, TPG—Productivity Inc. states:

> "Lean is a concept, a methodology, a way of working; it's any activity that reduces the waste inherent in any business process."

1.2 LEAN MANUFACTURING AND LEAN MAINTENANCE

The people at Toyota did not coin the term *Lean*. A research group at MIT coined it when they set about to analyze and define the Toyota Production System. The group, led by James P. Womack, published a book in 1990 titled *The Machine That Changed the World*, in which the term *Lean Manufacturing* was first seen in print. That book also put forth most of the processes that today define Lean thinking and the Lean Enterprise. In 1996, James Womack and Dan Jones, from that same research group, published *Lean Thinking*, which refined and distilled many of the principles put forth in the first book.

Today, James P. Womack, Ph.D., is the president and founder of the Lean Enterprise Institute (LEI). Based in Brookline, MA, LEI is a non-profit training, publishing and research organization founded in August 1997 to develop tools for implementing Lean production and Lean thinking, based initially on the Toyota Production System and now extended to an entire Lean Business System.

1.2.1 Elements of Lean Manufacturing

Until recently, manufacturers in North America, who have embraced the principles of Lean Manufacturing, did so without any measure of standardization. As a result, the face of Lean Manufacturing has many different looks. The Society of Automotive Engineers has recently published standards SAE J4000 (Identification and Measurement of Best Practice in Implementation of Lean Operation) and SAE J4001 (Implementation of Lean Operation User Manual). These standards are structured like a "5-S" type implementation, and are "high-level concept, low-level detail" documents. In the meantime and in spite of the lack of standards some characteristics are common to the majority of lean manufacturers:

- Waste Reduction
- Integrated Supply Chain
- Enhanced Customer Value
- Value Creating Organization

Unfortunately, many other characteristics essential for success have been often overlooked. Many plants undertaking the Lean approach seem to instill fear in their employees almost immediately. Their interpretation of Lean is that productivity must be increased using the fewest possible employees. To them Lean means a lean work force, one that will be achieved through faultfinding, blame and resulting layoffs. Thus "Lean" has acquired a very threatening meaning and is neither well received nor energetically supported within the workforce. This view of Lean Manufacturing is a common misconception. The characteristics that are most commonly left out of Lean Manufacturing implementations include:

- Committed Management
- Winning Employee Commitment
- Empowering Employees (Responsibility—and Accountability—at the Lowest Level)
- Optimized Equipment Reliability
- Measurement (Lean Performance) Systems
- Plant-Wide Lines of Communication
- All Processes and Workflows Defined
- Making and Sustaining Cultural Change
- Team Based Organization
- Continuous Improvement Practiced in All Departments and at All Levels
- Flatter Organizational Structure (less middle-level management)
- Measures of Performance Used
- Balanced Production (not maximum and not overproduction)
- Quality the First Time and Every Time

While these often omitted Lean implementation characteristics may not have the same visibility or promotion as the commonly included characteristics, they are every bit as important. Without them, any Lean transformation is ultimately doomed to failure. Needless to say, the largest problem in Lean Manufacturing is the failure to address a proactive reliability or maintenance process. Therefore, as waste is eliminated from the production process, and equipment operating time increases, reliability issues also begin to increase.

1.2.1.1 Lean Thinking and the Lean Organization

Some companies too often rely on production volume as their ultimate test for success. Lean isn't about productivity, and that's hard for many manufacturers to accept. It's about removing waste from the manufacturing process and building quality.

All of the business processes of a manufacturing plant must have common goals in the Lean transformation—gaining the competitive advantage. It means timely billing just as much as it means skilled, accurate machining. It means efficient sales and advertising just as much as it means reliable production equipment. All departments working together to relay information and data, to identify and correct problems and to maintain safe work spaces on the shop floor as well as in the business offices is essential for success of the Lean operation.

Factory workers must trust the company before they will put their hearts into improving—and possibly eliminating—their job routines. A fundamental rule of Lean manufacturing holds that a worker who is rendered unnecessary as a result of efficiency gains cannot be laid off. The company must guarantee employment in order to get the workers' full cooperation in identifying efficiency gains.

Lean is a comprehensive package that includes reducing inventory, standardizing work routines, improving processes, empowering workers to make decisions about quality, soliciting worker ideas, proofing for mistakes, applying just-in-time delivery and using a Lean supply chain. One might work without the others, but not for long. Lean thinking is elemental to a Lean transformation.

Transforming a manufacturing operation into a Lean enterprise requires an enormous transformation of the organizational culture. Lean thinking applies to the entire organization. Lean transformation requires a synergistic relationship between every branch of the organizational tree. Making gains in one department at the cost of efficiency in another department is definitely not Lean thinking. Lean transformation requires management commitment, job security and abolition of restrictive job classifications.

Lean Transformation is a journey, not a destination. Sustainment of the continuous improvement characteristic requires dedicated, committed leadership. It requires continual training and upgrade of skills, to include broadening those skills to cover diverse and non-restrictive job tasks.

1.2.1.2 The Role of Maintenance

LAWS OF MANUFACTURING MAINTENANCE

- Properly maintained manufacturing equipment makes many, quality products
- Improperly maintained manufacturing equipment makes fewer products of questionable quality
- Inoperable equipment makes no products

FACT 1

A manufacturing facility that has embraced all of the doctrines of Lean Manufacturing can't assume that it is equipped to implement those same "Lean practices" in the maintenance organization. In spite of expertise in Lean Manufacturing Practices, the unique requirements of the effective maintenance function call for a completely separate set of prerequisites.

FACT 2

Conversely, without a Lean Maintenance operation, Lean Manufacturing can never achieve the best possible attributes of "Lean." By definition, Lean means quality and value at the least possible cost. Without maximum equipment reliability—a product of optimized Lean maintenance practices—maximum product quality can never be attained.

A manufacturing plant with intentions of implementing Lean Manufacturing should begin with a few essential preparations. One of the most important preparations is the configuration of the maintenance organization to facilitate, first—Lean Maintenance, and next—Lean Manufacturing (see Figure 1-2).

The five principles of Lean implementation—Specify (value), Map (value stream), Apply (flow), Selectivity (pull) and Continuous Improvement (perfection)—are impossible to optimize in the maintenance organization without first understanding the foundation elements of successful maintenance and optimizing them before approaching Lean Maintenance (see Figure 1-3).

The very foundation of Lean Maintenance is Total Productive Maintenance (TPM). TPM should be established and operating effectively

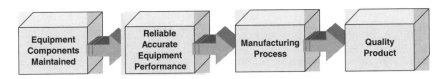

Figure 1-2 Lean Manufacturing

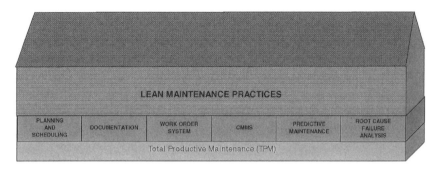

Figure 1-3 Lean Maintenance Practices

prior to applying the tools of "Lean." Without the foundation, you will be laying the bricks of Lean Maintenance on bare earth—the structure is destined to fail. Attempting to implement TPM and Lean simultaneously is akin to preparing for the Super Bowl while recruiting the football team. Recruiting a championship caliber team and honing their team effectiveness skills is obviously necessary before qualifying and preparing for the Super Bowl. Just as obvious should be making sure your maintenance operation is characterized as being of championship caliber as a team operation before preparing for and executing Lean maintenance practices. Adopting a proactive Total Productive Maintenance style of operation hones the skills of the maintenance and operations group to facilitate the Lean transformation (see Figure 1-4).

1.3 GOVERNING PRINCIPLES: WHAT IS LEAN AND WHAT IS NOT

1.3.1 What Lean Manufacturing Isn't

Above all, Lean Transformation is not double talk for downsizing. It means reassigning people and resources from useless work to value-adding work. Layoffs lead to defensive posturing; efforts by workers to limit production efficiency and prevent further improvements that will make their jobs unnecessary.

Positive employee reaction to Lean is crucial to success, and must be gained at the very beginning of the transformation. The key concepts of the Lean organization are teamwork, employee involvement, continuous improvement, communication and self-direction, all the key elements of cultural change. But unlike the failed "activity-based" programs of the 1990s, this is "on demand" cultural change. The need for it is obvious, even

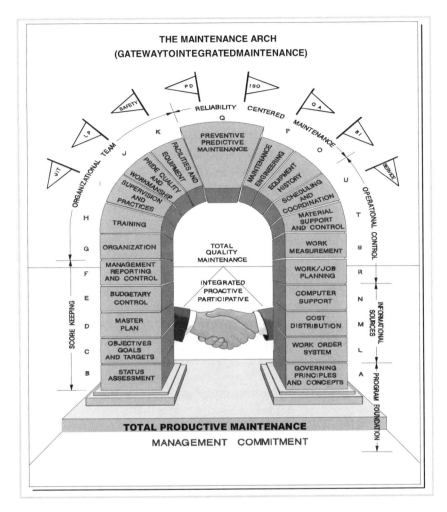

THE MAINTENANCE ARCH
(GATEWAYTOINTEGRATEDMAINTENANCE)

Figure 1-4 The Maintenance Arch (Gateway to Integrated Maintenance)

pressing. It is immediately applicable to supporting change on the shop floor. In the Lean Organization, staff positions and management levels are reduced, authority and responsibility are driven down to the lowest level, barriers fall and communication at all levels improves.

Even without the threat mentality and potential for job losses, the fast paced change of this magnitude during Lean transformation is stressful. Broader duties, steadier work pace, shift reassignment, more responsibility and new emphasis on flexibility and teamwork can lead to resistance. Increasing pay to compensate for the new demands is often justified and will aid in the transformation. Stressing the many positive aspects, such as better ergonomics, more variety, higher job satisfaction, job security and more input into improvements in safety, methods, equipment layout, tools, etc., also helps in the transformation to lean.

1.3.2 What Lean Manufacturing Is

A Definition:

Lean Manufacturing is the practice of eliminating waste in every area of production including customer relations (sales, delivery, billing, service and product satisfaction), product design, supplier networks, production flow, maintenance, engineering, quality assurance and factory management. Its goal is to utilize less human effort, less inventory, less time to respond to customer demand, less time to develop products and less space to produce top quality products in the most efficient and economical manner possible.

1.4 RELATIONSHIPS IN THE LEAN ENVIRONMENT

Lean is about waste reduction and customer focus. It's also about quality the first time and continuous improvement and it's about problem solving. But, perhaps above all, it's about people. Unlike traditional manufacturing organizations, people are not the problem in a lean enterprise; they are the problem solvers. Who knows more about problems with a step in the manufacturing process, the shop-floor operator or the middle level manager in his office filling out forms? And who is more likely to know what the solution is to the problem with a step in the manufacturing process? Lean empowers the shop-floor operator, encourages his involvement in waste reduction, customer relations and product quality, and in continuous improvement.

1.4.1 Information Integration in the Lean Organization

In order to function effectively in the lean manufacturing environment, the shop-floor operator needs to know more about customer needs, equipment maintenance and reliability and the supply chain; in general he needs to know more about the business operation. This need to know more about the business operation applies to everyone working in the Lean enterprise. Effectively integrating knowledge among organizational elements requires establishing communication systems that:

- Identify critical design issues as early as possible
- Encourage on-the-spot decision making using fewest resources to resolve critical design issues
- Promote knowledge sharing between hourly workers, management, design and interdepartmental
- Drive behavior of internal operations
- Drive behavior of suppliers and customers
- Accept formal as well as informal communication methods

Organizational awareness "across the board" is critical to both the early success of Lean transformations as well as to the long-term sustainment of Lean thinking. The very initial phase of the Lean transformation, following the decision to proceed, needs to be one of education. A hierarchy of progressively more informative presentations, dependent on their degree of involvement in the Lean transformation, should be prepared and provided to all plant employees. Presentations should not only describe the Lean transformation processes but should also provide senior management's vision and objectives, define the time line, describe the effect of the transformation on employees and EMPHASIZE job security aspects of the transformation. This kind of organizational awareness must also become an ongoing aspect of the Lean organization.

All of the business processes of a manufacturing or process plant must have a common goal for the Lean transformation—gaining the competitive advantage.

1.5 SUMMARY OF LEAN CONCEPTS

When considering the Lean Enterprise (see Figure 1-5), defining it must be done within several self-contained domains. There are first the

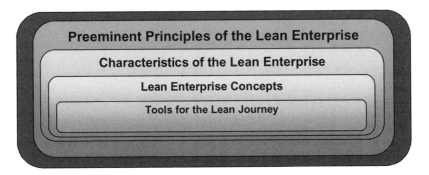

Figure 1-5 Lean Enterprise

preeminent principles of Lean that must be dominant in all aspects of the Lean Enterprises practices (including the subset of Lean implementation principles). Within that envelope are the characteristics of the operation, the concepts under which the enterprise operates and the tools used in making the lean journey.

Preeminent Principles

- Customer Focused
- Doing More with Less (Waste Elimination)
- Quality at the Source

Principles of Implementation—a Subset of Preeminent Principles

- Specify (value)
- Map (process/value stream)
- Apply (process flow)
- Selectivity (pull)
- Continuous Improvement (perfection)

Characteristics

- Standardize-Do-Check-Act (SDCA) to Plan-Do-Check-Act (PDCA)
- Next Production Line Process is Your Customer
- Quality the First Time, Every Time
- Market-in vs. Product-out
- Upstream Leveled Management Structure
- Let Data Speak
- Variability Control and Recurrence Prevention

Concepts

- Waste Reduction
- Integrated Supply Chain
- Enhanced Customer Value
- Value Creating Organization
- Committed Management
- Winning Employee Commitment/Empowering Employees
- Optimized Equipment Reliability
- Measurement (Lean Performance) Systems
- Plant-Wide Lines of Communication
- Making and Sustaining Cultural Change

Tools

- 5-S Process
- Seven Deadly Wastes
- Standardized Work Flow (TAKT Time)
- Value Stream
- Kanban (Pull System & Visual Cues)
- Jidoka (Quality at the Source)
- Poka-Yoke (Mistake [Error] Proofing)
- JIT (Just-in-Time)

2

Goals and Objectives

Definitions:

Objectives describe a desired state of organizational being. They are the accomplishments sought by an organization over a long period of time. A time span in excess of one year designates a course of action or a plan as long-range. A strategic course of action (plan) spanning several years has long-range objectives. Objectives are qualitative as well as quantitative.

Goals are quantitative, ultimate and strategic long-range aims. Properly selected, they can be motivating as well as productive tools. In a team environment, the team should set goals. Goals must, while being ultimate and strategic, also be attainable.

Targets are milestones to be reached in progressing toward goals and objectives. They are short range and are time and achievement related.

2.1 THE PRIMARY GOALS AND OBJECTIVES OF MANUFACTURING

It stands to reason that the primary objective of any business, manufacturing or otherwise, is to make money, usually in the form of profits. The size of the profit margin may be an objective or a goal. In the very largest of businesses, the profit margin may not be important, as long as it's positive, because making money is oriented around volume. But, in the smallest businesses, profit margin is everything because volume is low.

Another objective of most manufacturers is market share. To illustrate: Acme's primary objective is still to make money, but it's a longer-range objective, and it's much more money. Looking at the longer-range objective of industry dominance, the year one and two goals of 15% and 30% market share are elements of Acme's strategic plan for achieving their objectives of industry domination and making money. Each year Acme must measure their market share to determine their progress in executing their strategic plan.

Now that you thoroughly understand the relationships between objectives, goals and plans, let's take a closer look at more typical goals of most manufacturers. In actuality they will involve an interwoven network of relationships among the various business processes within the manufacturing organization.

2.1.1 Sales

Sales are right up there in relative importance among manufacturing goals. However, sales success depends on having a competitive advantage. Without a competitive advantage, only the very quickest of the fast-talkers in the sales staff will have much success. What makes up the competitive advantage? Here are a few of the more important aspects:

Aspects of Competitive Advantage	Responsible Organizational Element
• Price	
• Supply Chain Purchasing Raw Materials	**Top Management**

MRO Materials

- Customer Satisfaction:
  - Product Quality — Production/Maintenance
  - On-Time Delivery — Distribution
  - Fluctuating Demand Flexibility Production/Supply
  - Service:
    - Warranty and Repair — Service
    - Billing (Timely and Accurate) — Accounting
    - Personal Relationships — Sales & Service

Just this small sampling of competitive advantage aspects essential to successful sales illustrates how virtually every branch of the organization tree has some impact on the competitive advantage.

2.1.2 Production

Production levels are also high on the list of important manufacturing goals. Unfortunately, the trend in goal setting is towards maximizing production. This often results in excess product inventory from over-production. With more inventory than sales, the excess has to be stock-piled and incurs the added costs of storage—space, environment control, security, etc. With these additional costs we've just driven up price (or reduced profit margin) hence we've lost part of that competitive advantage. The goals work against each other.

Production level goals should be to match production to sales. When increased sales goals are met, production levels must be capable of rising to the increased sales level. Thus a plant's capacity must have the ability to deal with fluctuating production demands.

2.1.3 The Manufacturing Budget

Often, a major goal in manufacturing is to reduce the total cost per unit produced to a designated value. The total cost per unit produced or cost per unit (CPU) is equal to the number of items or units manufactured in a defined period of time divided by all of the plant's costs (expenses) during that same period of time. (Before Lean Thinking, many companies used labor cost per unit as a measure of performance—which ignored inflated costs of flawed processes.) For example, a company manufactures only PerfectWidget at Plant No. 1. In a single year Plant No.1 manufactures 12,000 PerfectWidgets. The expense portion of the budget of Plant No.1 for that same year totals $1,500,000. Then . . .

$$CPU = \$1{,}500{,}000 \div 12{,}000 = \$125$$

or, it costs Plant No. 1 $125 to produce one PerfectWidget. The price that the company can sell PerfectWidget for determines its profit margin. If PerfectWidget sells for $200, the profit margin is . . .

$$\text{Profit margin} = \text{Sales price per unit} - \text{CPU, or } \$200 - \$125$$
$$= \$75 \text{ Profit Margin}$$

Profit margin is more commonly expressed either as a ratio, of profit to sales, or as a percentage. For this example we would have . . .

$$\text{Profit} \div \text{Sales} = \$75 \div \$200 = 0.375\!:\!1 \text{ or } 37.5\%$$

We could expand the discussion to differentiate gross profit from net profit (margin), however for our purposes the definition provided above is sufficient for understanding follow-on discussions. It is important to note that profit margin is a key indicator of a business's health when trended over time. Temporary decreases in profit margin may be due, for example, to an increase in the cost of raw materials. However, longer term, steadily declining profit margins may indicate a margin squeeze suggesting the need for productivity improvements or a reduction in expenses.

2.1.3.1 Budget Elements

Table 2-1 is an abbreviated manufacturing company budget. The individual line items in a budget can be seemingly endless, so we have chosen to show a few typical basic budget elements to provide an understanding of the relative costs of various items and to convey a 'feel' for how a manufacturing company's budget is constructed.

2.1.3.2 Controlling Costs

In general, plant budgets are established for three different purposes:

- Allocating direct and indirect costs to products
- Providing a base for the annual profit/operating plan
- Internal cost control and performance evaluation at the functional/operating level

Within maintenance management, it is the latter purpose to which our focus is drawn. Firm control of expenditures is essential to the success of any individual operation. However, near/short-term control must not be achieved to the detriment of long-term success of the manufacturing/ maintenance operation. Accordingly, management requires reliable procedures and relevant records to determine:

- Where and why costs have been incurred historically
- How essentially and effectively managed historical costs have been
- The cost control effectiveness of each function
- How effectively the application of authorized resources has supported the broad organizational vision and mission
- What changes are anticipated that will influence future resource needs and how significant will that influence be
- Discretionary budget items that can be released or deleted as conditions throughout the budgetary period dictate
- Appropriate budgets reflective of the above considerations

Table 2-1
Abbreviated Manufacturing Company Budget

ACME MANUFACTURING January–March 2004 BUDGET		
Expenses	**Budget**	**Actual**
Direct Costs		
Direct Labor	$1,350,000.00	$1,285,780.00
Direct Travel	$67,500.00	$52,350.00
Direct Material (production)	$850,000.00	$857,250.00
Direct Parts/Consumables	$30,000.00	$36,300.00
Total Direct Costs	$2,297,500.00	$2,231,680.00
Indirect (Overhead) Costs		
Mortgage Service	$270,000.00	$270,000.00
Capital Equipment	$10,000.00	$9,950.00
Utilities	$36,000.00	$38,522.80
Insurance	$18,000.00	$18,000.00
Tax & Licenses	$6,300.00	$6,300.00
Sales and Marketing		
Salaries	$450,000.00	$450,000.00
Commissions	$120,000.00	$85,000.00
Advertising	$86,000.00	$86,000.00
Accounting Department		
Labor	$60,000.00	$60,000.00
Supplies	$5,000.00	$5,000.00
Postage	$2,400.00	$2,475.00
IT Department		
Salaries/Labor	$78,000.00	$78,000.00
License Fees	$12,000.00	$12,000.00
Hardware	$6,000.00	$8,300.00
Software	$6,000.00	$0.00
Other Departments		
Fringe Benefits, Etc.		
Total Indirect Costs	$1,165,700.00	$1,129,547.80
TOTAL EXPENSES	$3,463,200.00	$3,361,227.80
Revenue	**Budget**	**Actual**
Income From Production		
Sales Invoicing	$3,982,680.00	$3,445,258.50
Total Production Income	$3,982,680.00	$3,445,258.50
Income From Business Operations		
Favorable Legal Actions	$0.00	$100,000.00
Sale of Stock	$0.00	$128,055.00
Insurance Claims	$150,000.00	$150,000.00
Other Receivables	$30,000.00	$30,000.00
Total Business Ops Income	$180,000.00	$408,055.00
TOTAL REVENUE	$4,162,680.00	$3,853,313.50
TOTAL REVENUE MINUS EXPENSES	**$699,480.00**	**$492,085.70**

Production Maintenance Engineering Management (Floor) Quality Control

Includes: Straight Time Overtime and Salaried Labor

Variances generated by (material) market fluctuations, quantity discounts, failed receipt inspections, etc.

$3,445,258 – $3,361,227 = $84,031 Production Generated Income

Gross Income (Before Taxes)

- How actual costs compare to budgeted costs (this need is not only actuarial, but also on a dynamic basis so that variances can be avoided, controlled or limited as appropriate)
- Periodic results and progress
- Obstacles to the control process

A budget is understood to be a cost goal or an estimate of the cost of performing work at some future period for which a given level of support has been defined. A budget is not primarily historical, but should be considered always as a forecast of future expenditures (possibly greater, possibly less, than historical experience). Budgets will never be meaningful if simply established at an arbitrary percent increase over the last year. Nor will they ever be meaningful without extensive participation by the manager responsible.

A BUDGET IS NOT A LICENSE TO SPEND!

Cost distribution accumulates actual costs for comparison to established budgets. Sound analysis of historical cost data, adjusted for planned improvements and predicted changes in conditions, is the basis of realistic budgeting. Participation, by the responsible manager, in the budgetary process is essential if the budgets are to have meaning and are to be an effective tool relative to any of the three purposes listed previously, especially at the operating level.

In order to build effective maintenance budgets, it is necessary to build an effective work order breakdown structure and to have maintenance costs segregated by cost center and within cost center, by responsibility, equipment, work type (repair, alteration, preventive, reconditioning and capital versus expense) and by maintenance activity type (nature) (regulatory demand, safety and expansion versus continuing operations). Much of this segregation can only be realized through a well-conceived work order system and well-designed and well-defined maintenance and repair cost reporting procedures and guidelines because typical accounting systems are not this definitive.

For effective control, budgets must be consistent with responsibility. If an individual Cost Center Leader (position) controls certain expenditures, that Cost Center Leader (position) must share responsibility for budgeting and subsequently controlling these expenditures. No Cost Center Leader (position) should have authorization to approve work orders unless the related charges will be made to that Cost Center's (position's) budget, otherwise the Cost Center Leader's control is completely lost.

Like budget variances for other functions, maintenance variances can be analyzed in terms of volume and performance. Volume (or production) variances are largely influenced by the amount of support an equipment cost center requests—budget versus standard/estimate for work requested. Performance variances (standard/estimate versus actual charges for work requested) reflect the effectiveness at which maintenance support is provided. Variance analysis of this type is a very meaningful budgetary tool, but is also dependent upon meaningful work measurement to establish the standards/estimates. Segregation of budgets as volume variances by custodian has dramatic impact on control of emergency and urgent work demands and on support of maintenance scheduling since program benefits accrue to equipment cost center budgets as well as to maintenance budgets.

The cost of maintenance is neither a fixed nor a variable expense. It changes with volume in a stepwise pattern. This is because of the incremental nature of the staffing requirements of an effective maintenance department. Ironically, the maintenance staff percentage of total facility employment is likely to be largest, as high as 25%, in a small facility having up to 100 persons employed in operations. For 100 to 250 total operational employees, economies of scale drive the percentage down to about 12%. For larger facilities, the need for more specialization of the maintenance support staff brings the figure back up to the 18% range (see Figure 2-1).

Adequate control of expenditures is essential to the success of any program. A series of well-developed procedures and records employed to determine overall functional effectiveness, results, progress and obstacles to continual progress are required.

Sound analysis of historical data, tempered for planned improvements, results in realistic budgeting. Sufficient segregation of maintenance costs, by cost center and within cost center by responsibility (volume variances to requesting unit and performance variances to maintenance perform-

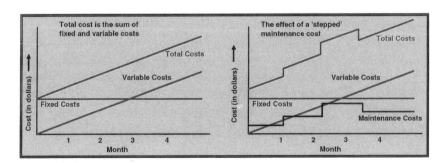

Figure 2-1 Cost of Maintenance

ing unit), by equipment, by type (repair, alteration, rearrangement, preventive maintenance, capital versus expense) and by cause is necessary. Such segregation is seldom available from the budgetary cost system, yet is required for meaningful control. It can be obtained through the work order system and CMMS (see Figure 2-2).

2.1.3.3 Optimizing Maintenance as a Cost Control Measure

Note that the partial budget previously shown has two columns—Budget and Actual. The goal here is to have actual come in at or below budget on expense items and at or above budget on revenue items. Incentives like bonuses and commissions are paid for outperforming the budget, with constraints such as meeting production and quality goals applied to ensure that ineffective corner cutting isn't used just to undercut a budget line-item number. Front-line managers are evaluated on their adherence to budgets; after all, as seen in the sample budget, cost overruns directly impact the company's income. Cost management on the shop floor often can involve attempts to control activities that are beyond local control.

The things, for example, that a maintenance manager can do to keep labor costs minimized are often at odds with his primary function. When a failure occurs, he can assign a mechanic who has the lowest hourly wage, or he can deny overtime and let production wait until morning before he sends someone to make repairs. Of course, keeping the equipment from failing would allow the maintenance manager to avoid ever having to "knowingly break his budget." Unfortunately the budget often tends to be a vertical "chain-of-command" enforcer. It reinforces top management's control and undermines empowerment of front line teams, which is essential to effective Lean practices.

So, should we just ignore cost control measures? The answer obviously is no, but how can we effectively control costs while adhering to Lean principles? Manufacturing companies must create a culture of thrift and continuous improvement, reinforced by a long-term, organization-wide reward system. The whole concept of Lean is based on recognizing which work adds value as well as identifying and eliminating non-value-adding work. The emphasis should be on managing value up rather than managing costs down. In the Lean organization of the future, it is just possible that budgets will become obsolete.

Cost Minimization in the maintenance operation is a matter of not performing unnecessary maintenance (increased labor costs, more off-line production time, etc.) and is also a matter of not missing required main-

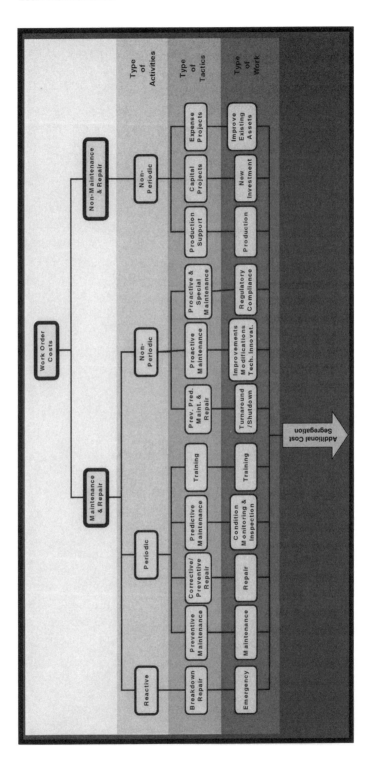

Figure 2-2 Work order costs

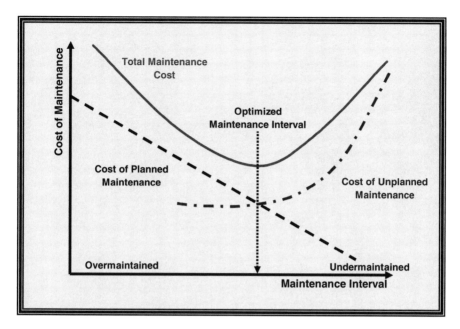

Figure 2-3 Cost Minimization

tenance (reduced equipment reliability, equipment failures, production downtime). Although this is a simple concept, achieving this balance is the basis of Lean maintenance. How to achieve this balance requires sound reliability engineering applied to a Total Productive Maintenance (TPM) system employing Predictive Maintenance (PdM) Techniques and Condition Monitoring. Utilizing elements of Reliability Centered Maintenance (RCM) is perhaps the best method to achieve this balance and balancing all of this to make sense out of optimum maintenance intervals is the secret to controlling maintenance costs (see Figure 2-3 above).

2.1.3.4 CPU—The Bottom Line

For corporate executives, net income after taxes—profit—is the bottom line. But for front-line managers and shop floor employees, that is a figure generally unknown and therefore unreal. The cost per unit produced, especially the portion of that cost that you are responsible for, is very real and readily measured. Tracking your cost reductions, tying them to production volume and displaying the results can provide instant feedback, and motivation, on the success of the Lean efforts that you have initiated. Table 2-2 is presented in order to provide some basis to measure of the cost of your maintenance operation.

Table 2-2
Financial (Maintenance Costs) Benchmarks

U.S. Industry (ALL) Statistics	Maintenance Cost as a % of CRV	Maintenance Cost as a % of Controllable Costs	Training Cost per Maintenance Hourly Employee	Maintenance Cost as a % of Total Sales	Contractor Cost as a % of Total Maintenance Cost
HIGH	6.0%	26.1%	1,990	14.1%	46.0%
MEAN	3.1%	18.0%	1,010	7.7%	21.0%
LOW	1.5%	8.4%	210	0.5%	0.0%
BEST: H (HIGH) OR L (LOW)	L[1]	L	H	L	L OR H[2]

Notes: 1. If a maintenance budget is below 2% of Current Replacement Value (CRV) the overall condition of the plant should be evaluated to determine if the infrastructure is undergoing slow degradation or if an excessive number of projects are required to maintain the plant equipment.
2. Either High or Low is acceptable depending on the strategic plans for the plant, i.e., if the decision has been made to out-source either a portion or all of the maintenance function.

2.1.4 Growth and Continuous Improvement

Lean Implementation begins by selecting a process for a Kaizen event—of no more than a week's duration. Although Kaizen literally translates as continuous improvement, in practice it means small incremental improvements. It is best done in little pieces, in the slivers of time that arise even in busy shops. This approach better integrates improvement with daily work, engages everybody rather than just a small team and fosters learning by everyday practice. Kaizen does not need to be an event. It does not even need to be perceptible. In successful Lean transformations, employees, when turned loose for improvement, have mixed daily work and improvement so skillfully that only close observation would reveal the small improvements occurring. This, in essence, is the magic of continuous improvement and the secret to sustaining improvement.

2.2 INTEGRATING LEAN GOALS WITH MAINTENANCE GOALS

2.2.1 Maintenance Objectives and Goals

It is important that you distinguish between the Maintenance Operation's Vision, its Mission and its goals and objectives. A vision statement should explain what the organization would like to become or where the organization would like to be in the future. On the other hand, a mission statement should explain the purpose of the Maintenance Operation's existence. Mission is the reason you do what you do.

Goals are the measurable steps toward fulfilling the mission while objectives are the organizational conditions that must be met in order to fulfill the mission. Maintenance objectives, goals and targets should be synchronized with the departmental mission statement and be consistent with the facility strategic and operations/production plans that are formulated to realize the company's vision. A vision statement is generated by the company ownership, or for publicly owned companies, by the Board of Directors and top-tier executives. Typical statements contained in vision statements are:

- To become the recognized leader, in terms of quality of product . . .
- To gain an identity as the leader in technical innovation of . . .

Leadership and Commitment

Maintenance Principle 1: The maintenance program must have management understanding, commitment, support and involvement—communicated via a well-conceived maintenance mission statement.

More powerful than the will to win is the courage to begin. Creating a better future starts with the ability to envision it. There is magic in a positive image. Top management can make it happen. If management is committed, it is amazing how people pick up the standard and carry it throughout the organization.

All policies and procedures established for any organization are based upon basic mission statements and specific objectives. Management vision and involvement is the solid substratum on which the maintenance arch is built.

An example of a typical maintenance mission statement:

> Our mission is to provide timely, quality and cost-effective service and technical guidance in support of short-range and long-range operating/production plans. We will ensure, through proactivity, rather than reactivity, that assets are maintained to support required levels of reliability, availability, output capacity, quality and customer service. This mission is to be fulfilled within a working environment, which fosters safety, high morale and job fulfillment for all members of the maintenance team while protecting the surrounding environment.

2.2.1.1 Maintenance Objectives

Maintenance Operation objectives must support both the plant's strategic and production plans and the plan's objectives. In addition they must support the maintenance operation's stated mission. The number one objective for all maintenance organizations everywhere is maintenance of equipment reliability. Additional maintenance objectives consistent with the mission statement are:

- Control maintenance workload by:
  - Maintaining the work backlog within prescribed limits by providing for forecasted resource requirements
  - Adhering to daily schedules
- Continually reduce equipment downtime and increase availability through the establishment of a preventive/predictive maintenance program (including failure analysis) designed, directed, monitored and continually enhanced by maintenance engineering

- Ensure that work is performed efficiently through organized planning, optimized material support and coordinated work execution
- Establish maintenance processes, procedures and best practices to achieve optimal response to emergency and urgent conditions
- Create and maintain measurements of maintenance performance
- Provide meaningful management reports to enhance control of maintenance operations
- Provide quality maintenance service in support of operational need

2.2.1.2 Maintenance Goals

It is essential to establish specific goals for achievement in relation to the plant's strategic and operational plans, both short- and long-term. Furthermore, targets must be set for maintenance performance in terms of equipment up time, maintenance costs, overtime, work-force productivity and supervisor's time at job sites. Such specific targets as these enable management to monitor progress and the effectiveness of the maintenance management program and to control activities by focusing corrective attention on performances or levels that consistently fall short of the targets.

Superficial goals lead to superficial results. A clear understanding of the mission is critical to success. A positive and productive mindset results in positive and productive performance. Build a climate, which helps others to motivate themselves. Always aim high and pursue those things that will make a difference rather than seeking the safe path of mediocrity.

Expect the best—it generates pride.

Instead of imposing goals on subordinates, allow team members to contribute to the goals setting process. People will feel they have control of their own environment and will raise their own standards. They will feel that accomplishment is its own reward and they are in control.

Everybody is motivated, but sometimes to work against, rather than for, the supervisor. Together, as a team, set clear and mutually agreed upon goals. People work harder to meet objectives, which they help to set. Clear goals lead to performance excellence. State the desired outcome in positive terms and be as specific as possible. Define goals so they are measurable. Set goals that are attainable and ensure that goals are relevant. Ask yourself whether they are economically viable.

Once goals have been established, it is imperative that everyone knows what they are. The most effective method is through tried and true advertising methods. Posters delineating the goals can be posted in conspicuous places where they will be sure to be seen by everyone. Maintain the posters. When they begin to show their age, or become damaged, replace them. It's just human nature to associate the "condition" of the media with the condition of the program. Keep them bright, clean, easily read and in good repair.

Develop the metrics process by which progress will be measured. This requires a system that makes goals simple to track so that you can reward or redirect at regular intervals. The following table illustrates such a measurable and easily tracked set of maintenance organization goals (see Table 2-3).

Table 2-3
Maintenance Goals

Indicator	Ultimate Goal
Backlog—Ready	2 to 4 weeks
Total	4 to 6 weeks**
Stores service (average)	95% to 98%
Materials delivered to job site	Above 65%
Stores turns per year	2 to 3
Preventive maintenance man-hours (includes PDM)	30% or greater
Unscheduled man-hours	Under 10%
On-the-job supervision	Above 65%
Schedule compliance	Above 90%
PM schedule compliance	Above 95%
Overtime	5%
On-the-job wrench time	Above 55%
% of planned work	Above 90%
Emergency maintenance labor hours	Under 2%
PPM routines/corrective WO (actions)	6:1 (without RCM)

* To be achieved by year-end.
** Excluding major overhaul (shutdown or outage).

2.3 THE NEED FOR, AND GAINING, COMMITMENT

"The productivity of work is not the responsibility of the worker but of the manager."

—Peter F. Drucker

The late Dr. W. Edwards Deming emphasized, "When quality is poor, blame the system, not the people, and management is the system." The same insight applies to maintenance. Others would say, hey, bad workers, but Deming said, no, bad system. He insisted on questioning the company's culture and management philosophy . . . telling clients that 85% of quality problems are the result of management errors. When a good performer is pitted against a poor system, the system wins almost every time.

Productivity gain requires a total commitment. Reforms do not work well in isolation. You must do a lot right if you want to make a quantum leap forward. It's not just the building blocks, but also how they are placed and held together.

2.3.1 The First Step: Top Level Management Buy-in

Executives in the most successful companies instill a passion for excellence in their entire organization. Executives lead these companies with a passion for excellence that pervades the business and creates an identity and focus for every employee. Senior executives at the top of these companies articulate consistent, direct messages that enable employees throughout the organization to understand how the company works, how performance is measured and how the company is organized around its core strategies.

Without this kind of top management enthusiasm for making the Lean transformation, the ripple effect of indifference will certainly kill the effort before it has even begun to show any improvements. Managers up and down the line take their cues from their own immediate bosses. For the maintenance operation this begins with a changeover from reactive maintenance to the proactive approach of TPM. Convincing top management of the gains, in terms of return on investment (ROI), that will be realized with the implementation of TPM and gaining its firm commitment to the process is a very necessary first step in the entire Lean transformation.

2.3.1.1 The Good

Proactive Maintenance Characteristics Control over the Maintenance Resources—With the advent of correct maintenance planning and scheduling procedures there is often a vast and rapid change in the understanding of what is required of the maintenance resources from week to week. This often can easily extend to monthly planning periods.

Increased Inventory Control—The dual effects of increased equipment reliability and better planning and scheduling will lead directly to increased control over the operation of the maintenance stores.

Elimination of much of the "Waste" of the business processes—With accurate planning and scheduling processes, much of the waste in the processes will cease to exist. Waste appears generally in the form of waiting times for materials, equipment availability and in the provision of inaccurate information.

Increased Accuracy in Maintenance Budgeting—With the increases in equipment reliability, large gains in budget accuracy are immediately possible. The ability to forecast maintenance requirements, either by equipment or activity, are vastly enhanced when we reach the planned stage of maintenance.

Reduced Maintenance Costs—In conservative terms a task that has been planned and scheduled is at least 50% more efficient in terms of both costs and time to complete. Using this as a standard and applying it to the amount of tasks that are now executed in an unplanned fashion we can easily see the range of savings that are possible. Proactive TPM combined with a Maintenance Excellence initiative (see Chapter 3, Section 3.1.2) has been documented to produce a Return on Investment (ROI) of as much as a 10:1 within three years. In addition, a proactive TPM organization that has adopted the principles of Maintenance Excellence will spend approximately 2% of the site's estimated replacement value annually in maintenance labor, materials, subcontracts, spare parts and overhead. This 2% target has been proven to be achievable. It is not uncommon for organizations to achieve at least 30–50% reduction in maintenance spending within 3–5 years. However, capacity increases and total production cost per unit decreases should be realized within the first year.

2.3.1.2 The Bad and the Ugly

Reactive Maintenance Characteristics Low Equipment Reliability (MTBF—Mean Time Between Failure—measured by dividing time by number of breakdowns or emergency work orders)—When MTBF is not measured one may look only at the "squeaky wheel" (problems you face day to day) and not necessarily the biggest reliability issue. Frequent breakdowns can be a drain not only on production capacity, but also on maintenance resources.

Low Mean Time to Repair (MTTR)—This indicator can often be very misleading as to the performance of the plant equipment as a whole. In a reactive state it is often very low. This is because the workforce is accustomed to having to repair equipment and to do so in a very fast manner. Although a positive, in terms of workforce abilities, it often indicates a situation in which the plant itself is often failing.

Inaccurate Inventory Planning—One of the significant effects of low equipment reliability is the inability of maintenance stores to accurately control the level of inventory required. When they cannot be sure what will be required tomorrow it is impossible to construct anything like a long range plan for managing the inventory levels in a satisfactory manner.

Many Uncontrolled Stores—An additional effect of poor inventory planning is the number of uncontrolled or personal stores that maintenance departments are inclined to keep. This is due to the fact that maintenance has no confidence in the stores department to adequately maintain the levels of stock required and stems from the poor equipment reliability.

Highly Reactive Workforce—With the effects of all of the factors above, the workforce in this situation is generally extremely reactive in nature. When trying to change the corporate culture of an organization, this can often be one of the most difficult areas to change. The workforce takes a great deal of pride in its abilities to keep the plant running and rightly so. There is a tendency to want to run off and "save the day."

2.3.2 Selling at Each Level

How is change that yields progress initiated? It starts with awareness and education that change is required. "You don't know what you don't know" may be a cliché, but it is one that is profoundly accurate. Attaining the competitive advantage required to meet the challenges we face today certainly requires constant scrutiny of emerging technologies and new thinking, but much of the problem is neither with new technology or new thinking. It is merely lack of understanding of the subtle influences that our current thinking and work processes have on our business. Many executive-level managers focus solely on production. Production results are certainly the shortest link to profitability, but such a narrow focus is the downfall of many businesses. Like root-cause analysis, everything affecting both production and cost-per-unit produced must be examined. Successful and profitable businesses understand the value of total integration of all the business functions.

Maintenance is a production enabler; in other words, maintenance helps to determine what percentage of capacity (full run output) can be produced, which in turn is utilized by production to define product output level. Reactive maintenance practices cannot yield production levels at desired capacity. Only proactive maintenance, maintenance that prevents failure and preserves operational production assets, can deliver the near capacity production levels required to sustain our businesses.

A key ingredient that cannot be overemphasized for successful implementation of change, whether to proactive TPM or to a Lean Enterprise, is that support and participation at all levels is absolutely essential. Upper management does not usually carry out steps in an action plan; they merely drive it and remove the roadblocks that get in the way and ensure that the plan and its execution are conforming to the company's mission and its short- and long-range objectives. Management commitment, or rather the lack of it, is the single most common reason for failure to fully achieve expected results. Only when management is (1) fully committed to the change and to creating the environment to allow and promote change to occur, and (2) dedicated to its successful completion, can the organization succeed in a fully realized implementation of progressive and meaningful change. Leadership at every level is critical and prerequisite for sustaining change and is often the missing ingredient.

2.4 MEASURING PROGRESS

2.4.1 Metrics

We often refer to metrics, which is just a term meaning "to measure" (either a process or a result). The combinations of several metrics yield indicators, which serve to highlight some condition or highlight a question that we need an answer to. Key Performance Indicators (KPIs) combine several metrics and indicators to yield an assessment of critical or key processes. KPIs for maintenance effectiveness have been discussed, defined and refined for as long as proactive maintenance has been around. KPIs combine key metrics and indicators to measure maintenance performance in many areas.

We need to be able to define where we are headed as a corporation, in regards to our maintenance management goals, and define the KPIs that we will need to monitor in order to reach our eventual goals. This process is unique to each corporation and needs to be developed independently. One of the more interesting points here is that KPIs can be created in a hierarchical and interlinked fashion, which allows management to pinpoint the root causes of system failures. In order to determine maintenance strengths and weaknesses, KPIs should be broken down into those areas for which you need to know the performance levels. In maintenance these are areas such as preventive maintenance, materials management process, planning and scheduling, and so on. Depending on KPI values we classify them as either leading or lagging indicators.

Leading indicators are indicators that measure performance before a problem arises. To illustrate this, think of key performance indicators as yourself driving a car down a road. As you drive, you deviate from the driving lane and veer onto the shoulder of the road, the tires running over the "out of lane" indicators (typically a rough or "corrugated" section of pavement at the side of the road that serves to alert you to return to the driving lane before you veer completely off the pavement onto the shoulder of the road). These "out of lane" indicators are the KPI that you are approaching a critical condition or problem. Your action is to correct your steering to bring your car back into the driving lane before you go off the road (proactive condition).

If you did not have the indicators on the pavement edge, you would not be alerted to the impending crisis and you could veer so far out of the driving lane that you end up in the ditch. The condition of your car, sharply listing on the slope of the ditch, is a lagging indicator. Now you must call a wrecker to get you out of the ditch (reactive condition). Lagging indicators (such as your budget), yield reliability issues, which will result in capacity issues.

A manager must know if his department is squarely in the driving lane and that everything is under control, as long as possible before it approaches and goes into the ditch. A list of some of the key performance indicators of the leading variety is illustrated in the Key Performance Indicators table (Table 2-4). Note that some of these indicators could be both leading and lagging when combined with and applied to other KPIs (Key Performance Indicators). (see Table 2-4)

Table 2-4
Key Performance Indicators

Reliability/Maintainability
- ➤ MTBF (mean time between failures) by total operation and by area and then by equipment.
- ➤ MTTR (mean time to repair) maintainability of individual equipment.
- ➤ MTBR (mean time between repairs) equals MTBF minus MTTR.
- ➤ OEE (overall equipment effectiveness) Availability × Efficiency (slow speed) × Quality (all as a percentage).

Preventive Maintenance (includes predictive maintenance)
- ➤ PPM labor hrs. divided by Emergency labor hrs.
- ➤ PPM WOs (work orders) #s divided by CM (corrective maintenance, planned/scheduled work) WOs as a result of PM inspections.

Planning and Scheduling
- ➤ Planned/Schedule Compliance—(all maintenance labor hours for all work must be covered and not by "blanket work orders"). This a percentage of all labor hours actually completed to schedule divided by the total maintenance labor hours.
- ➤ Planned work—a % of total labor hours planned divided by total labor hours scheduled.

Materials Management
- ➤ Stores Service Level (% of stock outs)—Times a person comes to check out a part and receives a stock part divided by the number of times a person comes to the storeroom to check out a stocked part and the part is not available.
- ➤ Inventory Accuracy as a percentage.

Skills Training (NOTE: A manager must notify maintenance craft personnel about the measurement of success of skills training.)
- ➤ MTBF.
- ➤ Parts Usage—this is based on a specific area of training such as bearings.

Maintenance Supervision
- ➤ Maintenance Control—a % of unplanned labor hours divided by total labor hours.
- ➤ Crew efficiency—a % of the actual hours completed on scheduled work divided by the estimated time.
- ➤ Work Order (WO) Discipline—the % of labor accounted for on WOs.

Work Process Productivity
- ➤ Maintenance costs divided by net asset value.
- ➤ Total cost per unit produced.
- ➤ Overtime hours as % of total labor hours.

NOTE: KPIs must answer questions that you as a manager ask in order to control your maintenance process. Listed is a sampling of recommended KPIs. They are listed by the areas in which a maintenance manager must ask questions.

2.4.2 Selecting Performance Indicators and Key Performance Indicators

The course to truly improve maintenance for the long term is definitely not an easy one. However, from the initial difficult period the system begins to manage itself and the snowball of continuous improvement starts to be propelled under its own momentum.

We will explore several levels and forms of maintenance productivity measurement. All functions and levels of management require control measures, but it is imperative that we understand the differences in need, precision, source and application.

Leading indicators are task-specific metrics that respond more quickly than lagging indicators (or results metrics). They are generally selected to anticipate progress toward long-term objectives that may not change quickly in response to effort. As an example, training should result in improved skills, which should result in improved equipment reliability. Implementing a precision shaft-alignment training effort might require years to affect overall reliability measures such as MTBF and MTBR. For this case a leading indicator of grouped MTBFs or MTBRs for applicable equipment might be selected to gauge the effectiveness of the training effort.

Technical metrics are used to measure the effectiveness of equipment management programs and systems at the plant, unit or equipment level. These metrics demonstrate the technical results of programs such as vibration monitoring, fluid (lubricating oil) analysis and thermography as the first step in gauging their contribution to corporate and plant performance. Technical metrics capture, in objective terms, results that can be trended over time to track progress toward program objectives and demonstrate improvement. For example, a metric may be the percentage of predictive procedures performed within one week of schedule.

Selecting and Applying Metrics—The value of meaningful metrics cannot be overstated; the impact of metrics that are inaccurate or inapplicable cannot be understated. Metrics must connect to the organizational objectives. All of the key processes should have one or more metrics to indicate goal compliance and progress. In each case the process owners and implementers must be involved in selecting metrics. The value of an effective CMMS or EAM is in its ability to retrieve automatic, real-time data that you can use. The Equipment Management process is directed toward adding value. Metric selection and reporting must be consistent with that principle.

There are several rules to follow in applying metrics:

- Good metrics focus activities on maximum benefits and value added
- Poor metrics lead away from optimum activities, often to unintended results
- Whenever possible, metrics should be positive, rather than negative (e.g., measure first-run quality, not rework)
- Avoid conflicting metrics
- Always examine complementary metrics together (i.e., there isn't much benefit in directing efforts to increase yield if quality is significantly below objective)
- Noncompliance with a metric should be followed by efforts to identify cause, full cost, and other effects of noncompliance; many organizations use Pareto analyses for this purpose
- Metrics must be used and kept current; metrics that are not regularly used should be eliminated

The more commonly used maintenance metrics or performance indicators can be classified into three categories:

- Measures of equipment performance, such as availability, reliability and overall equipment effectiveness
- Measures of cost performance, such as labor and material costs of maintenance
- Measures of process performance, such as the ratio of planned and unplanned work, schedule compliance

Typically, these performance indicators are tracked because of the following reasons:

- These indicators have been used by the organization in the past.
- Some of them are used for benchmarking with other organizations.
- The required data is easy to collect.
- Some of them are mandated by regulators or the corporate office.

These are diagnostic measures that determine whether the various aspects of maintenance operations remain in control or compare favorably with counterparts elsewhere. Thus they are used largely to support operational control and benchmarking purposes. In general, these generic measures are inappropriate to provide a holistic assessment of maintenance's performance. Additionally, they do not provide information suitable for predicting the plant's ability to create future value needed to support the business success of the organization. To achieve that end, performance measures that are linked to the strategy of the maintenance function must be tracked. These are known as strategic measures. Figure 2-4 illustrates a process for managing maintenance performance from the strategic perspective.

Figure 2-4 Strategic Maintenance Performance Management Process

Table 2-5
Balanced Scorecard Template

Strategic Objectives	Performance Measures	Targets	Action Plans	Perspective
				Financial
				Customer
				Internal Processes
				Learning & Growth

A core feature of the process is the Balanced Scorecard (BSC) that provides a balanced presentation of strategic performance measures around four perspectives: financial, customer, internal processes and learning and growth. Managers often find the strategy too abstract to guide them in making day-to-day decisions. By using the Balanced Scorecard, the strategy is translated into something more understandable and readily acted on—long-term (strategic) objectives that relate performance measures and their targets and action plans. Table 2-5 above is a Balanced Scorecard template.

The BSC is a powerful communication tool for providing a sharp focus on factors that are important to maintenance in making contributions to business success of the company. It enables complete and balanced assessment of unit performance and guards against sub-optimization because all the key measures that collectively determine the total performance of maintenance are monitored.

The BSC or performance scorecard is a performance measurement system that helps a plant pursue its key success factors. The scorecard uses both internal and external benchmarking and employs a relevant cascading method of performance goal setting. Achievements are acknowledged and celebrated on a "real-time" basis and not at the traditional annual review.

For a balanced scorecard process to be motivational it must provide timely and accurate data. Simplicity is a key to the validity of measurements and the tractability of problems to their root cause. Data collection design must employ simple and easy-to-maintain databases to ensure data integrity. When people are trained in this process and are permitted to participate in relevant goal setting, Performance Management can motivate teams to higher achievements—including the exceeding of growth and profit expectations.

Five key elements of the balanced scorecard process are:

1. Establish a "no status-quo" mindset—if you're not winning, you're losing
2. Define company "key success factors"—examples: cost, speed and quality
3. Identify stretch goals that are relevant to the company's "key success factors"
4. Implement training/coaching programs—education is the pathway to excellence
5. Celebrate each goal achievement and raise the bar—don't wait until next year

For a mature performance management process, "benchmarking" has become the standard for establishing performance objectives. Benchmarking is still one of the most ill-defined management concepts and is one of those words that mean different things to different people. Our preferred definition comes from Xerox, who describes benchmarking as: "the continuous process of measuring our products, services and business practices against the toughest competition and those companies recognized as industry leaders."

The objective of benchmarking is to build on the ideas of others to improve future performance. The expectation being that by comparing your processes to best practice, major improvements can be realized. You should not consider carrying out external benchmarking until you have thoroughly analyzed your internal operations and an effective system of internal measurement has been established.

So what kind of results can you expect when a management team introduces the process of the balanced scorecard? First, people will become motivated and focused on the continuous improvement of their company's critical success factors. Second, personal and team achievements will become recognized and rewarded—creating an exciting, winning, work environment. Teamwork will improve and employee retention will rise. Finally, and most important, is the company wide euphoria as "bottom line" results improve and financial pressures no longer create a stressful and defensive work environment.

1. Strategic Measures

These are measures applied by executive management to specify and monitor how well each function must perform in support of the competitiveness and ultimate viability of the enterprise. Internal attainment versus external industry leaders and world-class benchmarks is particularly applicable to Strategic Measurement. (see Table 2-6) Such measures for maintenance include:

Table 2-6
Strategic Measurement

Measure	Essential Data Sources	Goal
Maintenance Cost Ratio to:		
Sales Dollar	Accounting/10 K	<6%
Total Units/Volume Produced	Accounting	Trend
Total Manufacturing Cost	Accounting	12 to 14%
Total Asset Value—Gross	Accounting/10 K	Var. 6 to 7%
Total Asset Value—Net	Accounting/10 K	11 to 12%
Equipment Replacement Cost	Accounting/10 K	2%
Million Gross Occupied Sq. Ft. per Year	Accounting/Engineering	Var.
Combined Equipment, Buildings and Grounds Maintenance Cost Per Million Gross Occupied Sq. Ft. per Year (Maintenance, Utilities, Energy, Housekeeping and Grounds)	Accounting	Var.
Investment Maintained per Mechanic	Accounting	>$5 M
Contractor Cost as Percent of Total Maintenance $'s	Accounting	20–35%
Number of Maintenance Crafts	Contract	4 or Less

Table 2-7
Measures Applied by Controller to Monitor Operating Results and Budgetary Performance

Measure	Essential Data Sources	Goal
Maintenance Budget Variance Analysis By:		
Total	Accounting	Minimum
Primary Account	Accounting	Minimum
Responsibility:		
—Supervisor	Work Order System	Minimum
—Custodian	Work Order System	Minimum
—Volume vs. Performance	Work Order System	Minimum

Table 2-8
Measures Related to Regulatory Compliance, Employee Safety, Utility Expense, Sanitation, and Working Environment

Measure	Essential Data Sources	Goal
OSHA Injuries per Million Hours of	Personnel/Payroll Records	<5%
Maintenance Payroll	Accounting	
Total Payroll	Accounting	
Utility Expense per Million Gross Occupied Sq. Ft. per Year	Accounting/Engineering	Var.
Kilowatt Hours per Million Gross Occupied Sq. Ft. per Year	Accounting/Engineering	Var.
Custodial Expense per Million Gross Occupied Sq. Ft. per Year	Accounting/Engineering	Var.
Grounds Maintenance Expense per Million Gross Occupied Sq Ft per Year	Accounting/Engineering	Var.
Grounds Maintenance Expense per Acre per Year	Accounting/Engineering	Var.

2. Financial and Budgetary Control Measures
3. Process & Facility Integrity
4. Internal Customer Service
5. Asset/Equipment Reliability

Table 2-9
Indices Measuring How Well Maintenance is Meeting
Needs of the Internal Customer

Measure	Essential Data Sources	Goal
Facility Condition Audits: Ongoing and Comparative	Staff Effort	
Customer Satisfaction		
Response Time to Urgent Requests	Shop Floor Data Collection	96% < 15 min.
Overdue Work Orders	Accounting/Engineering	<5% min.
Complaint Level	Accounting/Engineering	<2% min.
Satisfaction Surveys	Staff Effort	>98% min.
Satisfaction Relative to Cost	Composite Analysis	Trend
Recent Repair Jobs Requiring Crew Return (Job not done right the first time)	Work Order System	<5% min.
Backlog Weeks		
Ready Backlog	Work Order System	2 to 4 wks
Total Backlog	Work Order System	4 to 6 wks
Backlog Status	Work Order System	Var.

6. System Integrity
7. Organizational Integrity
8. Operational Control (Intrinsic Maintenance Efficiency)
9. Material Support
10. Trending and Predictive Maintenance

Table 2-10
Primary Measures Related to Maintenance Quality

Measure	Essential Data Sources	Goal
Maintenance Downtime by Equipment (Uptime)		
As Percent of Scheduled Run Time	Shop Floor Data Collection	0.5 to 2.0%
Mean Time Between Failures	Equipment History	Increasing Trend
Percent of Failures Addressed by Root Cause Analysis	Equipment History	>75%
PPM Schedule Compliance	PPM Schedule of the Work Order System	>95%
PPM Delinquency by Weeks Delinquent	PPM Schedule of the Work Order System	<5% past due 1 to 4 wks.
Ratio of Corrective Work Orders to PPM Inspections	Work Order System	1:6*
Call-In Frequency	Payroll System	Trend
Maintenance Overtime Percentage	Payroll System	5 to 15%*
Annual Training Per Mechanic, measured in:	Personnel Records	
Hours		>100/yr.
Dollars		$2 to $3K/yr.
Percent of Maintenance Labor Budget		5 to 10%*
Average Number of PMs per Thousand Occupied Sq. Ft. per Year	Engineering Calculation from Work Order System Data	Var.
Average PM Hours per Thousand Sq. Ft. per Year	Engineering Calculations from Work Order System Data	Var.

* Indicates Generally Accepted Industrial Best Practice Benchmarks.
These are the primary measures related to maintenance quality. They also impact internal as well as External Customer Satisfaction and might, therefore, also be added to the preceding list.

Table 2-11
Measurement Indices of Supportive Control System
Administration

Measure	Essential Data Sources	Goal
Qualitative Program Assessment Improvement Potential	Baseline and Periodic	0.800*
Percent of Total Maintenance Hours Covered by Work Order by:	Work Order System	
Comprehensive Work Orders		50%
Standing or Blanket Work Orders		Minimal
Percent of Total Paid Maintenance Hours Captured by Work Order Charges	Work Order System	100%
Urgent Response Work Orders by Requestor Highlighting those to be inappropriate "Service Calls" (i.e., within spec or tolerance)	Work Order System	Act & Trend
Percent of Work Orders Covered by Estimate/Percent of Work Orders where actual hours differ from estimated hours by more than 15%	Work Order System	90%
Percent of Work Orders Covered by Planned Job Packages	Work Order System	80%

* Indicates Generally Accepted Industrial Best Practice Benchmarks.

These indices measure how effectively the supportive control systems are being administered. High system integrity is essential if any of the other benchmarks are to be realized.

Table 2-12
Measures of Organization/Staffing Relative to Maintenance
Mission

Measure	Essential Data Sources	Goal
Percent of Crew consumed by:	Work Order System	
Urgent Response		10%
PPM Inspection		30%
Relief of Planned Backlog		60%
Mechanics Per	Calculation	
First Line Supervisor		8 to 15
Planner		20 to 30
Maintenance Engineer		40 to 70
Clerk (Administrator)		20 to 50
Storeroom Attendant		25 to 40
Support Person (Composite)		5
Absenteeism	Payroll	<3%
Average Work Orders per Million Gross Occupied Sq. Ft. per Year	Work Order System	Var.
Average Work Order Hours per Million Gross Occupied Sq. Ft. per Year	Work Order System	Var.

These measures indicate how realistic are the organizational structure, table of organization and staffing levels relative to the Maintenance Mission.

Table 2-13
Maintenance Organization Measures for Monitoring Program

Measure	Essential Data Sources	Goal
Sampling of Work Force Activity (Crews and Supervisors)	Work Order System	
Percent of Direct work (Wrench Time)		55 to 65%
Percent Supervisory Time-At-The-Job-Site		65%
Performance Against Standards (Std./Act as %)	Work Order System	
Crew Efficiency—Ultimately the Best Use to Build Up and Substantiate Staffing—Required to Provide a Specific Level of Support/Excellence		>85%
Operational Measures (Trended)	Work Order System	
Schedule Compliance (actual sch. hrs completed/total hours available)		>90%
Ratio of Planned to Unplanned Labor—Hours		9:1
Percent Planned Work		90 to 97%
Material Dollars Installed Per Labor Dollar Expended	Work Order System and Accounting	$1.00 to $1.50*
Custodial Expense per Custodian	Accounting	Var.
Maintenance Improvement (Comparison of Current to Past and Base Years)	Composite Analysis	>7% per yr

These are the measures used by the maintenance organization itself to know where the program currently is at and to ensure continuous functional improvement. Focus on the trend. Build upon incremental improvement.

Table 2-14
Measures of Purchasing/Stores Support of Internal Customer

Measure	Essential Data Sources	Goal
Storeroom Service Level	Stores Requisition Module	>95%
Vendor Performance	Purchasing Records	
On Time Delivery		>98%
Material/Part Performance		>98%
Materials Delivered to the Job Site	Stores Requisition Module	>50%
Percent Storeroom Dollar per Equipment Replacement Value	Accounting	<.5–.75%
Stores Turns per Year	Stores Requisition Module	3 to 5
Stores Disbursements as Percent of Total Maintenance Materials	Stores Requisition Module	<30%

These measures evaluate how well Purchasing and Stores are supporting the internal customer (Maintenance).

The following indicators also are recommended for consideration for long-term trending as maintenance program management tools:

- Equipment (by classification) percentage out-of-service time for repair maintenance
- Mean time between equipment overhauls and replacement
- Number of vibration-related problems found and corrected per month
- Number of vibration-related work orders open at the end of the month
- Number of vibration-related work orders over 3 months old
- Number of problems found by other Predictive Maintenance (PdM) techniques (i.e., infrared thermography, ultrasonics, lube oil analysis, etc.) and corrected per month, work orders open at the end of the month and work orders over 3 months old
- A monthly record of the accumulated economic benefits or cost avoidance for the various PdM techniques
- Number of spare parts eliminated from inventory as the result of the PdM program
- Number of overdue PM work orders at the end of the month (total number of PM actions should decrease)
- Aggregate vibration alert and alarm levels (trending down)

2.4.3 Maintain and Publish the Track

Support and enthusiasm for change can quickly wane without a process of continuous reinforcement. One of the most effective measures is an updated (at least monthly) tracking chart of progress that is displayed publicly. The display location should be visible to the entire plant, not just the maintenance staff. Although employees often downplay, at least verbally, the significance of change progress, they actually take much pride in their successful progress.

Tracking charts needn't track every individual performance indicator—that would lead to hundreds of charts—but can track overall progress for general categories of performance measures. For example:

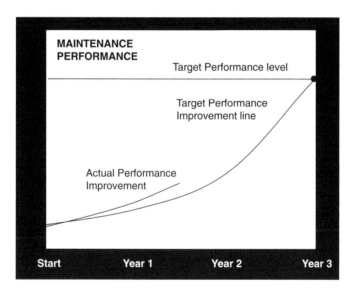

Figure 2-5 Maintenance Performance

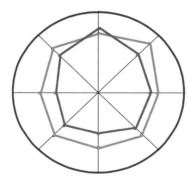

Figure 2-6 Radar or Spider Chart

This tracking chart combines the first thirteen indicators of the Operational Control category #8 of Maintenance Productivity Measures. The measures may be combined in any manner that makes sense to you, but they should be listed so that the contents are clearly understood.

The Radar or Spider Chart is also an effective tool for illustrating performance. The outer circle represents World Class Performance, the black plot is last quarter's performance and the gray is this quarter's performance. A dramatic illustration of the improvements made, it also readily shows where performance in one area (at the top) has declined.

3

Total Productive
Maintenance (TPM)

3.1 TPM (FINE-TUNED) IS LEAN MAINTENANCE

The very foundation of Lean Maintenance is Total Productive Maintenance (TPM). TPM is an initiative for optimizing the reliability and effectiveness of manufacturing equipment. TPM is team-based, proactive maintenance and involves every level and function in the organization, from top executives to the shop floor. TPM addresses the entire production system life cycle and builds a solid, shop-floor-based system to prevent all losses. TPM objectives include the elimination of all accidents, defects and breakdowns.

3.1.1 Elements and Characteristics

TPM is not a short-lived, problem-solving, maintenance cost reduction program. It is a process that changes corporate culture and permanently improves and maintains the overall effectiveness of equipment through active involvement of operators and all other members of the organization. TPM requires sponsorship and commitment from top management in order to be effective.

Most organizations that implemented TPM, failed to achieve the results that were anticipated. TPM was seen as a cost-cutting venture and was never sponsored or committed to by upper management.

The required TPM investment, as well as the return, is very high. Over time, the cooperative effort creates job enrichment and pride, which dramatically increases productivity and quality, optimizes equipment life cycle cost and broadens the base of every employee's knowledge and skill. TPM cannot be applied to unreliable equipment, therefore the company must initially bear the additional expense of restoring equipment to its proper condition and educating personnel about the equipment.

Team activities are basic to TPM. Teams at top management, middle management and shop floor levels carry out TPM activities. Each type of team has its own objectives and part to play.

Safety is a cornerstone of TPM. The basic principle behind TPM safety activities is to address dangerous conditions and behavior before they cause accidents. Workplace organization and discipline, regular inspections and servicing and standardization of work procedures are the three basic principles of safety. All are essential elements in creating a safe workplace.

The Eleven Major Losses

TPM activities should focus on results. One of the fundamental measures used in TPM is Overall Equipment Effectiveness or OEE.

$$OEE = \text{Equipment Availability} \times \text{Performance Efficiency} \times \text{Rate of Quality}$$

World-class levels of OEE start at 85% based on the following values:

90% (Equipment Availability) × 95% (Performance Efficiency) × 99% (Rate of Quality) = 84.6% OEE

The OEE calculation factors in the major losses that TPM seeks to eliminate.

The first focus of TPM should be on major equipment effectiveness losses, because this is where the largest gains can be realized in the shortest time. There are 11 major areas of loss, and they fall within four broad categories:

Planned-shutdown losses
    1. No production, breaks, and/or shift changes
    2. Planned maintenance
Downtime Losses
    3. Equipment failure or breakdowns
    4. Setups and changeovers
    5. Tooling or part changes
    6. Start-up and adjustment
Performance efficiency losses
    7. Minor stops (less than six minutes)
    8. Reduced speed or cycle time

Quality Losses
9. Scrap product/output
10. Defects or rework
11. Yield or process transition losses

Planned Shutdown Losses
Valuable operating time is lost when no production is planned. However, there are reasons that this is not often considered a "loss." It is probably best stated as a "hidden capacity to produce."

> Loss #1: Production is not scheduled. The facility may be a three-shift operation where the third shift is the maintenance shift. It may be a plant that doesn't run on Saturdays and Sundays. Employee shift changes, breaks, or meal times also fall into this category. During breaks, some equipment gets shut down. If an operation is a continuous process, obviously, that doesn't happen. But in plants making piece parts, there are assembly lines and processes that shut down for breaks and shift changes.
>
> Loss #2: Planned maintenance includes periodic shutdowns of equipment, processes and utilities for major maintenance. These shutdowns represent a period of time when no production occurs. Typically, OEE calculations factor planned maintenance out of the equation. It is assumed that it's planned maintenance—you have to do it, you can't reduce it, you can't eliminate it, so leave it in. Auto racing represents a different story about "planned maintenance." A planned-maintenance stop or "pit stop" on the NASCAR circuit in 1950 was typically near four minutes long. Now pit crews are performing the same pit stops in 17.5 to 22.5 seconds. What's happening in that "less-than-20-second" period are the same things that were happening in 1950 in four minutes: change four tires, dispense 22 gallons of fuel, make chassis adjustments, wipe off the windshield, give the driver some water and wipe the rubber dust off the radiator. In about 20 seconds, the car is back on the track. At some point, somebody said, "We can do these pit stops in less than four minutes, can't we?" Planned maintenance is a pit stop. TPM advocates must ask, "How much time are we spending in the pits?"
>
> A walk-through of almost any plant uncovers ways that the same amount of maintenance can be done in less time or that more maintenance can be done in the same amount of time. And those things can be accomplished not with more contractors and not with more work hours, but just by doing things differently and working smarter, not working harder.

Downtime Losses
Downtime is the second category of major equipment losses. This category includes the following:

Loss #3: Typically, when a company's personnel consider losses, they think of equipment failures or breakdowns. But there are other unplanned downtime losses.

Loss #4: This loss includes how long it takes to set up for production processing and how long it takes to shift from one product or lot to another. Determining these losses should take into account how long it takes to start up after a changeover and run a new product. In auto racing, this type of loss includes the preparation and setup for qualifying and racing.

Loss #5: This loss includes the time it takes to make tool changes and production-part changes. Industry in which certain tooling, machine devices or parts have to be changed can learn from an auto racing pit stop. The techniques are the same. In an actual pit stop, every single action is taken into account.

Loss #6: This loss occurs when equipment or processes are started up. It includes the warm-up time and the run-in time that must be set aside to get everything in the process ready to produce quality output. Pit stops are in part successful if the driver optimizes the car's speed on the slowdown lap before the pit stop and the speed-up lap after the pit stop. If the driver brings the car up to speed too quickly, the drive train and tires may be damaged. A race car handles poorly until the new tires are properly conditioned on the track.

Performance Efficiency Losses

The third category of major equipment losses is performance loss, when machines operate at less than designed speed, capacity or output.

Loss #7: These types of loss—minor stops or "machine hiccups"—are the little things that companies usually don't track. Quite often, equipment downtime of less than six minutes is not tracked. However, consider the impact if this six-minute downtime occurs during each shift in a three-shift, five-day operation. That all adds up to $1\frac{1}{2}$ hours of downtime per work week or 75 hours of lost production per 50-week year. Little losses add up.

Loss #8: Equipment and processes running at less than design speeds and cycle times result in lower output. As machines age and components wear, they tend to run slower. At times, machines are run at lower speeds because the people who operate and maintain them have compensated for problems and believe that running them slower is better and results in fewer breakdowns.

Quality Losses

The final category of major equipment losses is loss of quality. One of the fundamental truths of TPM is this:

If equipment is available 24 hours a day, 7 days a week . . . and if it's performing at its highest design cycle rate . . . then

if it's not producing the highest level of quality,
it's just producing scrap at full capacity.

So, quality is a very important element among the major losses.

Loss #9: Scrap loss is fairly straightforward. Every time equipment runs and produces unusable product or output, valuable operating time is lost.

Loss #10: Defective output, even if it can be reworked or recycled, is considered a major loss to be eliminated. As with scrap, every time equipment runs and produces unusable product or output, valuable operating time is lost.

Loss #11: Yield or transition losses often occur when equipment and processes require a warm-up or run-in time. During that time, they often produce an off-quality output. This type of loss also includes the lost output that results from transitioning from one chemical product to another in a process. Equipment running and wasting raw materials create yield and transition losses.

TPM can yield results in two months—sometimes two weeks—when activities are focused on results and there is regular monitoring, recording and trending of OEE data. It is not as important to focus on OEE percentage, as it is to focus on each of the factors of OEE and perform root-cause failure analysis on the major losses of each—equipment availability, performance efficiency and rate of quality. Use the OEE data to communicate how well the equipment is performing and how well the TPM activities are working.

As the foundation of Lean Maintenance, TPM is popularly viewed as a Japanese concept. To be more accurate it should be viewed as a program that effectively integrates a number of maintenance management concepts that did not originate in Japan. Yet, to the credit of the Japanese, they have a knack for turning good ideas into enormously successful practices. The Japanese business culture is more receptive to TPM requirements than is the dominant business culture in the United States (see Table 3-1).

3.1.1.1 Organization

In organizing the maintenance function there are several basic structural considerations that should be followed. The following are universally recognized sound organizational principles:

Maintenance management should be structured level with production management.

**Table 3-1
TPM requirements, United States versus Japan**

Typical Conditions in Japan	Typical Conditions in the U.S.
Total corporate commitment to TPM	Lack of management involvement
Very long-range planning	Focus on quarterly results
Few cost constraints	Severe cost constraints
Pressure to succeed from top	Less sustained pressure from the top to succeed
Practically no limit on training	Limited training time, yet plenty of time for meetings
Ability to absorb concurrent activities	Inability to absorb concurrent activities
Employees volunteer own time	Time constraints

Maintenance is not subordinate to production; rather it is a supportive service vs. subordinate.

Regardless of the organization style used, there should always be a current and complete organizational chart that clearly defines all maintenance department reporting and control relationships, as well as any relationships to other departments. The organization should clearly show responsibility for the three basic maintenance responses: routine, emergency and backlog relief.

Interfaces between Production and Maintenance should be clear and divisions between roles, responsibilities and authorities should be well defined within the organizational structures. The Maintenance organization structure should recognize three distinct (separate but mutually supportive) functions, so that each basic function receives the primary attention required:

• Work Execution
• Planning and Scheduling
• Maintenance Engineering

a. Work Execution

In general, there are seven different patterns, which show up in organization of maintenance activities:

• Organization by craft
• Organization by area

- Organization within production department
- Combination of craft and area
- Contract maintenance—partial or total
- Organization by work type
- Combination maintenance and production teams by area

Craft Organization—Much maintenance work is done by specialized craftsmen whose unions enforce trade rules resulting in the organization of maintenance labor by crafts. Thus maintenance jobs are first broken down into craft elements (electrical, instrument, sheet metal, machinist, welder, pipe fitter, etc.) and then each element is assigned to the appropriate craftsmen under the supervision of a craft foreman who directs the work of one or two crafts covering the entire facility. Thus, craft organizations are centralized and are assigned to jobs throughout the facility by central scheduling or dispatching service.

A variation of the craft pattern of organization—the functional-craft type—is used by some multi-plant companies. Here each foreman is assigned a major responsibility (e.g., maintaining electrical equipment or buildings and grounds) and provided with a work force composed of the requisite craftsmen. Thus, the buildings and grounds unit includes millwrights, painters, masons, gardeners and carpenters. This functional-craft form of organization is based on craft skills, but recognizes the functional task of organizing and administering maintenance work.

Functional work does not lend itself to the area type of supervision, either because it requires specialized skills or because the nature of the work requires maximum mobility. The plant engineer will assign such work to one of the central craft supervisors.

Area Organization—The area concept of supervising and controlling the maintenance function derives its name from the use of relatively small maintenance areas in which the activities of assigned maintenance personnel are directed and controlled by one individual known as area supervisor for maintenance.

The maintenance function is decentralized and maintenance crews are scheduled or assigned to areas within the plant, building or group of plants or buildings. Each area foreman is responsible for maintaining uninterrupted production in the area. The craftsmen and craft groups assigned to the foreman have the required skills to carry normal workloads of the area. However, if additional craftsmen are required, the area foreman may requisition them from other areas.

Production Departmental Maintenance—This is the old historical structure before the evolution of maintenance as a separately managed discipline. It is still found in small organizations that cannot justify separate supervision for maintenance crews of less than six or so craftsmen.

Combination Craft and Area Organization—Under this type of organization, the central shop is expanded by subdividing the shop into a series of specialized crafts, thus increasing the number of craftsmen. Some craftsmen are permanently assigned to shops and others to areas to take care of minor repairs, adjustments and even construction work, so that production can continue without interruption. Some craft activity (pump and turbine overhauls, re-tubing, rigging, machine shop work on lathes and grinders, valve repair, etc.) is centralized as a shop or central craft function and work is performed out in the field under the function's supervision—not area supervision. (Electrical and instrument work is also a good example of functional work.) The central maintenance shop supervisor(s) is responsible for shop work, craft administration and field consultation as required to meet production demands.

Special project work is under the direct supervision of the special projects supervisor. The types of work designated as special projects are varied. The special projects supervisor may be assigned as a relief supervisor for another supervisor who has an overload of work or he may be assigned to execute some major or special project.

All fieldwork, except functional work and specifically assigned special project work, is under the direct supervision of an area supervisor. Within a given area, the area supervisor directs the activities of all craftsmen assigned to perform work except those reporting to a functional supervisor. The area maintenance supervisor is responsible for obtaining materials, special tools and equipment indeterminable by planning and other needs that will expedite the work in progress. Timekeeping and the distribution of hours to jobs for payroll, planning and accounting purposes also are delegated to the area maintenance supervisor.

The area maintenance supervisor is not held responsible for supervision of personnel performing functional work in this area, or working under a supervisor responsible for a special project affecting his area. However, he or she is responsible for bringing to the attention of the assigned functional or special project supervisor any instances he observes of inadequate or inappropriate methods, poor workmanship, improper conduct or behavior that in his opinion is detrimental to his area.

All of the work must be coordinated by planning with the field area foremen, craft or functional foremen and production to ensure timely compliance.

Contract Maintenance (Partial or Total)—Under this organization, maintenance work is left entirely up to contract maintenance forces. The plant may choose to keep certain crafts and turn all other work over to the contractor for supervision and work execution. The plant may elect to plan and schedule the contractor force or let it handle the function. In general though, outside contractors are employed to perform (1) work, (2) recurring non-emergency work and (3) peak load work during situations such as shutdowns, turnarounds, construction, jobs, etc. The decision to use outside forces is usually based on a study of both cost and intangible factors.

Organization by Work Type—Effective control of the maintenance function depends upon clear accountability for each type of demand placed upon the organization. The three principal types of demand are routine or preventive, emergency and planned work.

An organization can be structured to facilitate control of each type of work. Such a structure is composed of three major operating groups covering the three principal types of demand. The basic concept of this structure is the establishment of two minimally sized crews to meet the routine and emergency demands and a third group devoted to planned maintenance.

The routine or preventive maintenance group is responsible for the performance of all management-approved routine tasks in accordance with detailed schedules and established quality levels. Their work:

- Is specifically defined
- Performed according to a known schedule
- Performed in a planned pattern
- Involves a consistent work content
- Requires a predictable amount of time

The group is not interrupted by emergencies or backlog, thereby protecting the integrity of the preventive maintenance schedule.

The emergency group has the responsibility of handling essentially all emergency demands, using assistance only when necessary. This allows the planned maintenance group to apply its manpower to backlog relief.

The planned maintenance group is responsible for all work other than emergency and routine. The group is divided into two crews, one covering work performed primarily in the shops and the other covering work performed in the field.

b. Planning and Scheduling

The responsibilities of Maintenance Planning and Scheduling include:

- Customer liaison for non-emergency work
- Job plans and estimates
- Full day's work each day for each man (capacity scheduling)
- Work schedules by priority

- Coordinates availability of manpower, parts, materials, equipment in preparation for work execution
- Arranges for delivery of materials to job site
- Ensures even low priority jobs are accomplished
- Maintains records, indexes, charts
- Reports on performance versus goals

c. Maintenance Engineering

In general terms the function of Maintenance Engineering is the application of engineering methods and skills to the correction of equipment problems causing excessive production downtime and maintenance work. Their responsibilities are to:

- Ensure maintainability of new installations
- Identify and correct chronic and costly equipment problems
- Provide technical advice to maintenance and proprietors
- Design and monitor an effective and economically justified preventive maintenance program for the TPM program
- Ensure proper operation and care of equipment
- Establish a comprehensive lubrication program
- Perform inspections of adjustments, parts, parts replacements, overhauls, etc., for selected equipment
- Perform vibration and other predictive analyses
- Ensure equipment protection from environmental conditions
- Maintain and analyze equipment data and history records to predict maintenance needs (Selected elements of RCM—Reliability Centered Maintenance)

TYPICAL ORGANIZATION STRUCTURE (see Figure 3-1)

3.1.1.2 Work Flow and the Work Order

Work Flow Scheme—One element of the transition planning process that can be a major stumbling block is analyzing existing work flow patterns and devising the necessary workflow and organizational changes required to make use of a Computerized Maintenance Management System (CMMS). This process can be difficult for the employees involved. When workflow shifts from a reactive to the proactive posture of TPM, planned and scheduled maintenance will replace the predominantly corrective maintenance style. The CMMS will provide insights into organized, proactive work flow arrangements through its system modeling.

Although you can tailor workflow and organizational attributes to match your plant's requirements, they still must work within any con-

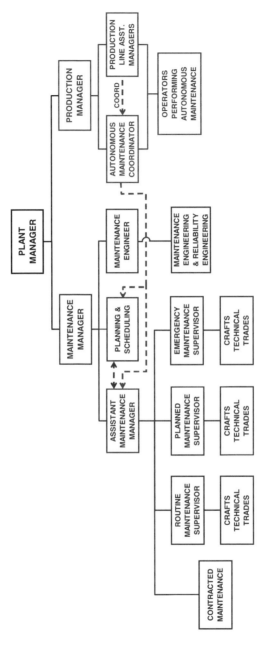

Figure 3-1 Planned Maintenance Group

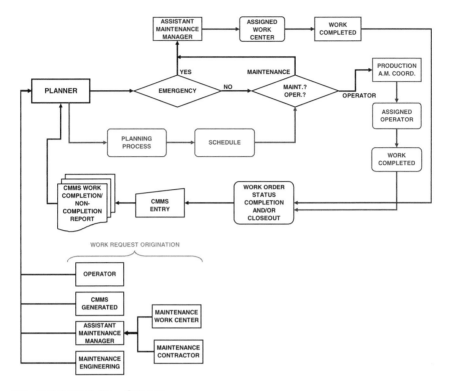

Figure 3-2 Work Flow Scheme

straints imposed by the CMMS. Of primary importance is keeping focused on the ultimate objective—a proactive Total Productive Maintenance organization that will assist in reaching the standards of best maintenance practices. (see Figure 3-2)

Work Request/Work Order Process Flow.

Work Order System—There is probably an existing work order system that is at least loosely followed. Again, the CMMS will help in defining changes to, or complete restructuring of, any existing work order system. The work order will be the backbone of the new proactive maintenance organization's work execution, information input and feedback from the CMMS. All work must be captured on a work order—8 hours on the job equals 8 hours on work orders.

The types of work orders an organization needs will need to be defined. They will include categories such as planned/scheduled, corrective, emergency, etc. The work order will be the primary tool for managing labor resources and measuring department effectiveness.

3.1.1.3 Support Functions

Virtually all departments in a manufacturing environment provide support for the Total Productive Maintenance function. There are major support roles for some:

Information Technology (IT) Department

- Technical Operation and Upkeep of CMMS
- Format and Content of Management Reports
- Archival (e.g., Equipment History) Maintenance
- CMMS Operator (User) Training

MRO Storeroom/Purchasing

- Timely supply of parts and consumables
  - Accurate reorder levels
  - JIT Suppliers
  - Pre-staging of material for scheduled maintenance
- Standardized material and consumables
- Standardized parts suppliers (no multiple suppliers of the same item)
- Maintain usage data (via CMMS)
- Purging of obsolete material

Production Department

- Operators perform equipment cleaning
- Part of "early warning" system
- Individuals who spot problems pitch in to restore equipment—resulting in less downtime.
- Trained and certified to perform designated routine maintenance tasks in their zone as the need arises—One Point Lessons (1 Per Week)
- Operators and mechanics are organized into zone teams

Major maintenance work, requiring high craft skills, is performed by centralized (or less centralized) maintenance forces.

3.1.2 Best Maintenance Practices and Maintenance Excellence

The most significant areas that directly affect production quality are maintenance and its success in sustaining equipment reliability (see Section 2.2.1.1). The potential financial gains from improving product quality can far exceed a company's maintenance budget. More often than not, management does not recognize this factor as well as the more obvious ones. The focus here is equipment reliability, viewed from both the quality of maintenance perspective as well as the quality of product perspective, and how it impacts business profitability.

Best Maintenance Practices (BMP) are established standards for the performance of industrial maintenance. Measuring a plant's existing maintenance process using the yardstick of BMP can reveal both the degree of maintenance impact on reliability and also permit identification of the specific maintenance processes causing variations in equipment reliability. Following are several examples of best maintenance practice standards. With each are listed possible causes for not meeting the standards and potential solutions for resolving maintenance process variations from these best maintenance practices (see Table 3-2).

This "Best Maintenance Practice" sampling, courtesy of John Day, former Engineering/Maintenance Manager at Alumax, acknowledged worldwide for over 20 years as the Best in Maintenance through achievement of recognition as a World Class operation.

The samples in Table 3-2 are provided to identify just a few areas of variation together with possible solutions. For a company to eliminate all of the variations in equipment reliability and the resulting lost revenue, it must approach, in a programmatic way, a fundamental change in the maintenance process and optimizing equipment reliability. A practical and effective approach to determine the need for, and implement, such a fundamental change should:

1. Identify whether an equipment reliability problem exists, and whether it impacts quality, and then determine the magnitude. Measure (in dollars of lost revenue) waste caused by equipment reliability issues.
2. Perform a maintenance assessment to identify where the variations are in the maintenance process. For example, are they in procedures relating to preventive maintenance (PM), planned maintenance schedules, the maintenance storeroom, emergency work, crew structure, or some other maintenance-related function?
3. Develop an action plan and timeline, together with benchmarks and performance metrics, to reduce variations in the maintenance process to achieve an acceptable quality level.
4. As you measure, implement necessary improvements to the maintenance process as required and continue to measure in order to gauge the success of the changes.

Companies that want to compete more effectively in today's marketplace must be progressive in accepting the need for, and implementing, change. Eliminating variations in the process and thus improving equipment reliability and therefore product quality can induce a snowball

Table 3-2
BEST MAINTENANCE PRACTICE STANDARDS (Sampling)

Measurement Standard	Possible Causes	Solutions
No "self-induced" equipment failures (Note: statistics show that 70% of equipment failures in industry today are "self-induced")	Lack of skilled work force	Skills assessment and training
	Operator errors	TPM/operator procedures
	Reactive culture	Change measurement
	Preventive maintenance procedures (PMs) not performed properly	PMs must be managed as an experiment
30% of all labor hours should be on PM	PMs not being performed to a standard	Have detailed procedures
	PMs not a high priority	Measure PM compliance
90% of all work orders come from preventive maintenance	PM inspections are turning into repair activities	Train personnel in proper PM execution
	90% of all maintenance work is not planned and scheduled	Implement a true planned/scheduled maintenance program
Emergency work is less than 2% of total maintenance labor hours	No PM schedule compliance	PM schedules must be completed within 10% of the frequency (e.g., 30-day frequency/PM compliance to within 3 days plus or minus)

Courtesy of John Day.

effect by providing increased market share, adding revenue (and profit) to the bottom line and increasing employee morale and effectiveness, thereby reducing costs associated with a maintenance program operating in a reactive mode. An up-to-date, proactive Total Productive Maintenance program can pay for itself through the elimination of product quality variations.

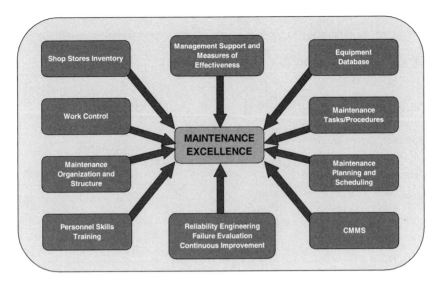

Figure 3-3 Factors to Consider for Maintenance Excellence

The Maintenance Excellence Maintenance Arch (see Chapter 1, Section 1.1.2.3) foretells the large payback that will be realized by achieving Maintenance Excellence. The return on investment (ROI) over three years is documented again and again to be near 10 to 1 and often even greater. Achieving Maintenance Excellence requires the integration of many factors that include attainment of Best Maintenance Practices, which are established by you to meet BMP Standards. Other factors essential to Maintenance Excellence are illustrated in Figure 3-3.

Maintenance Spending: World Class Maintenance benchmarking suggests that a proactive TPM organization that has adopted the principles of Maintenance Excellence will spend approximately 2% of the site's estimated replacement value annually in maintenance labor, materials, subcontracts, spare parts and overhead. This 2% target has been proven achievable. It is not uncommon for organizations to achieve at least 30–50% reduction in maintenance spending within 3–5 years, however, capacity increases and total production cost per unit decreases should be realized within the first year.

Production Volume Increases: In a proactive environment, the site will experience fewer equipment failures (and the time losses associated with them). These losses usually fall into the following categories:

- Unscheduled Downtime
- Off-quality production

Figure 3-4 Best Maintenance Practices Scale

- Missed production schedules
- Waste disposal
- Reduced throughput operation
- Startup/shutdown losses
- Customer claims/product returns
- Low labor and energy productivity

At the top end of Maintenance Excellence are those companies achieving recognition as World Class Maintenance Operations. Few companies will ever reach this status because it takes commitment from the top-level executives all the way to the shop floor and everyone in between. World Class status requires executing Best Maintenance Practices in all areas at the highest end of the scale. But it is achievable. Ask John Day. (See Table 3-2 and Figure 3-4)

3.1.3 Maintenance Skills Training & Qualification

The skill level of the maintenance personnel in most companies is well below what businesses and industry would categorize as acceptable. Lack of adequate skills is being reflected in reduced equipment reliability and increased maintenance costs. There is an answer to the shortage of adequately skilled labor. In an effective Total Planned Maintenance environment, a skills improvement and training program is structured to provide real value to the maintenance operation. An effective and properly developed and implemented maintenance skills training program can have dramatic impact on equipment reliability. But, the training must be focused on solving real problems to give results, as quickly as possible and at the same time it must meet a manufacturing plant's long-term goals.

Determine what the training is meant to accomplish. Performing a Job Task Analysis (JTA) will help you define the skill levels required of maintenance department employees to perform maintenance tasks on your plant's installed equipment. The JTA should be followed with a Skills Assessment (SA), using written exams and practical demonstrations, of employee knowledge and skill levels. Analyze the gap between required skills from the JTA and available skills from the SA, to determine the amount and level of training necessary to close the gap. Instituting a qualification and certification program that is set up to measure skills achievement through written exams and practical skills demonstration, together with maintenance effectiveness assessments (metrics) based on equipment reliability, will provide you with feedback on training effectiveness. It will also assist in resource allocation when scheduling planned/preventive maintenance tasks. The maintenance-engineering group is responsible for the continuous update/improvement of the skills training program.

TPM and Lean Practices both involve autonomous maintenance or operator performed maintenance during production. Operators can perform many of the "in-operation" maintenance requirements much more efficiently than maintenance staff, brought in by work orders, which might have to wait for the proper conditions to perform a particular maintenance requirement.

Including production line operators in the maintenance skills training loop therefore makes sense from efficiency point of view as well as from an equipment reliability standpoint. Just the reverse situation also has important implications for production capacity and quality viewpoints. A maintenance mechanic performing "off-line" maintenance can include setup or pre-setup calibration of production equipment as a closeout step of the maintenance requirement much more efficiently than can the operator, who may have to recreate equipment conditions all over again in order to complete his setup. Cross training of operators and maintenance personnel can produce substantial efficiencies in the manufacturing environment.

Multi-skilled maintenance technicians are becoming more and more valuable in modern manufacturing plants employing PLCs, PC Based Equipment and Process Control, Automated Testing, Remote Process Monitoring and Control and/or similar modern production systems. Maintenance Technicians who can test and operate these systems as well as make mechanical adjustments, calibrations and parts replacement obviate the need for multiple crafts in many maintenance tasks. The plant processes should determine the need for and advantages of including multiple skills training in the overall training plan.

Employing a qualification and certification program that is integrated with the training program can provide many benefits. It provides for continuous evaluation of skills available as well as skill deficiencies for fine-tuning and continuous improvement of the skills training program. By employing a combination of self-study, classroom training and on-the-job training/coaching—together with written examinations and practical skills demonstration—Maintenance Managers can identify precisely what skills their employees have mastered and match maintenance tasks and maintenance personnel more effectively. Skills that are lacking, or in need of improvement, will be readily identified and specific steps for upgrading skills will be immediately available to acquire the needed skills in minimum time. In addition, qualification and certification removes most of the guesswork from the promotion selection process.

A Maintenance Training and Qualification (MTQ) Program should include these primary features:

1. Plant specific Task/Subject Area (TSA) listing
2. Training and Qualification Summary (TQS) for each TSA and employee
3. Maintenance Training
   a) Self-study Assignments
   b) Classroom Instruction
   c) On-the-job Training
4. Skill Evaluations
   a) Written Examination
   b) Skill Execution/Demonstration
5. TQS Completion Record and Skill Area/Skill Level Certification

Refer to Appendix B for examples of documentation for these training, qualification and certification components.

Selected TQS Series define the path to position certification through defined qualification steps as illustrated below in Figures 3-5 and 3-6.

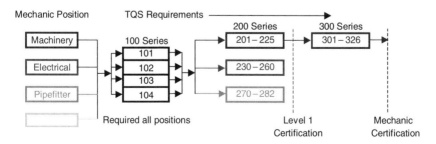

Figure 3-5 Selected TQS Series

No.	Subject Areas and Tasks	Type Training CR	SS	OJ	Mach.	Instr.	Elect.	Pipe.	Type Training
100	**General**								
101	General Knowledge & Skills		X		X	X	X		SS - Self Study
102	Mechanical (Basics & Safety)		X		X				CR - Classroom
103	Electrical (Basics & Safety)		X				X		OJ - On-the-job
etc.									
200	**Machinery**								
	Machinery (Basic)								
201	Centrifugal Pumps		X	X					
202	Reciprocating Pumps		X	X					
203	Piping & Valves		X	X					
204	Fans		X	X					
etc.									
	Machinery (Level 1)								
210	Bearing Types & Lubrication	X	X	X	X				
211	Fastener Types & Torquing	X	X	X	X				
etc.									

Figure 3-6 Typical Task/Subject Area—TSA (selected portions for illustration purposes)

3.1.4 MRO Storeroom

Spare Parts Inventory Reduction—Reactive maintenance organizations typically have a large inventory of spare parts even while, at the same time, engaging in excessive emergency purchasing activities to address breakdowns. As a business moves toward a proactive TPM culture with more planned and scheduled work, the need for maintenance is identified early enough to be able to order materials and receive them in a JIT scenario, before failure occurs. In addition, improved organization of the spare parts storage locations helps eliminate duplication of inventory and enables calculation of appropriate minimum and maximum stocking levels and economic order quantities. It also enables the identification of obsolete inventory that can be returned to the vendor or otherwise discarded.

A very useful tool of the Lean storeroom is management of inventory by ABC analysis. ABC analysis is the method of classifying items involved in a decision situation on the basis of their relative importance. Its classification may be on the basis of monetary value, availability of resources, variations in lead time, part criticality to the running of a facility, new customer parts unique to that product and others.

Cycle inventory can be managed through ABC analysis. Once ABC items are coded in the file maintenance system, they are sorted by the ABC code. The CMMS randomly selects *A* items so that all can be processed typically in a two-month period, *B* items typically can all be processed in a six-month period, and *C* items can all be processed in a year period. The daily count would reflect this percentage of parts. A review of this process should take place quarterly to ensure proper ABC codes are issued for the parts. Standard Costs are used to determine the cost of the part. A definition of Standard Cost is the normal expected cost of an operation, process or product, including labor, material and overhead charges. It is computed on the basis of past performance costs, estimates or work measurements.

Another use of ABC is in the management of storage areas by the storeroom manager. *A* items are needed to be more closely reviewed due to their dollar value and importance to the facility. Normally, *A* items are stored at the lower levels of the bin. *B* items are stored in the mid levels and are normally replacement parts that may not have the criticality or dollar value of the *A* items. *C* items are stored in all other areas.

Obsolescence budgeting also takes the management of ABC analysis into consideration. *A* items have the most impact on the budget, if it is determined to be obsolete and scrapped from inventory. *B* items are second, and *C* items are third. The mix of *ABC* affects a monthly budget for obsolescence. The storeroom manager or supervisor ensures the best use of the budget each month to scrap materials that have no added value to the storeroom inventory, with input from maintenance. These parts may be a component of equipment that is being replaced. (Each piece of equipment should have a bill of materials developed to identify all parts required to maintain a specific piece of equipment.) They can also be a part of equipment that is no longer going to be used in the next model year. In addition these may no longer be functional parts in the storeroom. Through slow-moving activity reports, parts analysis and visual methods of the storeroom parts, these can be identified. However, the danger to obsolescence is the fact that *A* parts may not have a use for several years (shows on the slow-moving activity report); but, due to its critical importance, the parts may be needed at a later date. Because of its criticality, the lack of these parts may shut down the plant. The slow moving activity report would not detect this need. Management and storeroom management need to consider all aspects of the parts before it is scrapped to obsolescence. ABC analysis provides a perspective that enhances this decision-making.

Still another use of ABC analysis is in the reorganization of the storeroom. Yearly, a review of parts storage areas needs to be made by the storeroom manager. In this analysis, ABC should be considered so that the *A* parts are continually being moved to the lower or easier access areas. Inventory adjustments to *A* items need to be reviewed more closely and investigated. Recounts typically occur when deviations occur. *A* parts should be located in an area where they can be visibly observed and controlled. Most of the time these parts should be accessible when needed. By constantly reviewing the storage of parts, the management can continually reorganize to the stores areas so that the best organization can be presented.

ABC analysis even affects lot-sizing considerations. A plant using EOQ (Economic Order Quantity where a fixed order quantity is established that minimizes the total of carrying and preparation costs under conditions of certainty and independent demand) uses ABC as well, so that inventory levels are minimized with the higher cost part.

The storeroom manager needs to be aware of the ABC analysis in all of his or her management techniques. It is an important tool in his decision-making for the binning of material, the counting of material on a daily basis, planning and scheduling with lot sizes and his obsolescence review. A world-class storeroom addresses these issues efficiently and effectively.

Companies transitioning to proactive maintenance have significantly reduced the amount of emergency purchasing and costly overnight shipping costs. Independent survey results suggest that inventory levels in maintenance operations that transition from reactive to proactive environments can expect at least a 17% reduction and an average of nearly a 50% reduction.

Inventory holding costs typically run between twenty and thirty percent of the inventory value on an annual basis. Reduction in spare parts inventory therefore results in an immediate and recurring cost savings to the business.

3.1.5 Planning and Scheduling

Planning is a staff function. As such it should be organizationally independent of the specific maintenance supervisor(s) it is supporting. The planning function should report to a level of maintenance management, which is at least one level above the first-line supervisory level (the level supported by planning on a day-to-day basis).

If there are more than three positions in the control organization (including planners, schedulers, material coordinators, clerks, dispatchers and maintenance engineers), they should report to a maintenance control group leader.

Planner/Scheduler Working Relationship—A conscious decision is necessary in regard to working liaisons. Is the planner going to interface directly with the operating department, or is that relationship to be a function of the maintenance management level to which planning reports?

Direct liaison—The planner is supporting maintenance management and supervision as well as the operating unit to a maximum degree. This is in keeping with the current team concepts and individual participation and involvement precepts.

Indirect liaison—The planner supports only the maintenance manager. He is, in effect, a staff assistant to the maintenance manager. All other liaison is accomplished within the line (both within maintenance and with operations).

Direct liaison is the preferred mode!

Internal Planning & Scheduling Structure (Horizontally across facility or vertically within function)—In either case, it should be centralized to preserve independence. Normally this is horizontal—with one planner/scheduler performing planning, scheduling, material coordination and operating liaison for all maintenance work associated with one or more areas, or for one or more central craft groups.

An alternative in large organizations is vertical segregation of planners, schedulers and material coordinators. The selection of structure is often predicated upon available skills. Planning requires more craft knowledge than scheduling. The latter structure can conserve planning capability. Similarly, when the material management system is complex and system knowledge is limited, the material coordinator position should be considered.

Factors Influencing Planner/Scheduler Control Span—Control span ratios should only be used as guidelines. Actual planner staffing depends upon a number of factors and should be tailored to fit the specific local situation, considering:

- Current state of maintenance management installation
- The planning and scheduling organizational structure
- Peripheral responsibilities placed upon the planners
- Other maintenance staff support in place (maintenance engineers, maintenance clerks, material coordinators, training coordinators, PM coordinators, relief supervisors, maintenance control manager)
- Complexity of craft structure
- Complexity of operating organization, which maintenance supports
- The number of craft personnel performing planned/scheduled work
- The level of planning and scheduling needed
- The method of estimating used
- The level of liaison and coordination for which the planner is responsible
- The current state of planner support system (labor and material libraries, etc. (see Figure 3-7)

	Points Assigned
Planning and Scheduling Structure *Separate from material coordinating (vertical structure) - 1 point* *Combined (horizontal structure) - 2 points*	
Number of Crafts Coordinated *One - 1 point* *Two - 2 points* *Three - 3 points* *Four - 4 points*	
Level of Planning *Craft and general description with schedule - 1 point* *Craft, general instructions, special tools and major materials with schedule - 3 points* *Craft, specific instructions, tools, materials, prints and schedule - 5 points* *All the above plus work methods described - 7 points*	
Level of Estimating *Estimates or historical data - 1 point* *Slotting against benchmarks or labor library - 3 points* *Analytical estimating - 5 points* *Measured time development of individual jobs - 7 points*	
Total Points	

Total Points		Ratio Craftsmen to Planner
4 to 7	3	0 : 1
8 to 12		25 : 1
13 to 17		20 : 1
18 to 22		15 : 1

Figure 3-7 Determination Worksheet ration of craftsmen to planners

3.1.6 CMMS (Computerized Management Maintenance System)

This TPM Program element is considered optional, but highly re-commended. It should be considered by all but the very smallest manufacturing plants. A properly implemented CMMS can produce management efficiencies that are virtually impossible to achieve without CMMS.

When considering the installation of CMMS, either as a new install or an upgrade from an existing install, it is important to define your expectations and needs from CMMS. During the first ten to fifteen years of CMMS development, the functions or modules available from the various vendors were relatively standardized. In recent years, CMMS vendors have developed a divergence from the standard by offering a continually expanding array of specialized functions. Don't be dazzled into adding features for which you haven't defined a need. Sometimes these specialty functions actually impede the effective utilization of the more important

CMMS functions. A survey of more than 600 maintenance departments[1] highlighted six key functions that are "must have" features in a CMMS:

- Work Order Management
- Planning Function
- Scheduling Function
- Budget/Cost Function
- Spares Management
- Key Performance Indicators (KPI)

Additionally, there are important features of a CMMS that can help to ensure the success of your installation. When selecting a CMMS vendor, evaluate their product for:

- Ease of Use
- Management Support
- Low Learning Curve
- A Defined Maintenance Work Process

Many companies have purchased a Computerized Maintenance Management System or Computer Managed Maintenance Systems (CMMS) or Computerized Asset Management System (CAMS) with the intent that the system will be the silver bullet that solves all the maintenance problems—and it can if it is properly implemented and its features effectively utilized. Computer Managed Maintenance Systems have become more sophisticated and much more capable over the last five to eight years, yet many CMMS users feel that their systems have failed to deliver the desired results. This is seldom, if ever, because of the CMMS capabilities. There are some fundamental issues to recognize in selecting, installing and implementing a CMMS. CMMS Software developers will install their system and train your staff to operate it efficiently. They will also define what kind of data is required to be entered into the various CMMS modules and data fields. When they've completed their installation and training, the software is still void of data. In order to obtain effective Maintenance Management Information from CMMS software the following are absolute requirements:

- All of Your Facility's Pertinent Data Must be Entered
- 100% Data Accuracy Is a Must
- Formats Must be Understandable to Both You and the Software

---

[1] CMMS Benchmarking Survey 2003: Reliabilityweb.com, Cmmscity.com and Maintenancebenchmarking.com

The realization of these three attributes of the data to be entered constitutes implementation of the CMMS Software.

Just as TPM is the foundation for Lean Maintenance, the completeness and accuracy of equipment information is the foundation for an effective CMMS implementation. The following must be addressed to ensure a functional CMMS is established in Table 3-3:

Table 3-3
CMMS Implementation

Fit the user needs	The ultimate user is the maintenance or production worker on the floor. Management is the user of the information, not the typical initiator of the information.
Be fully implemented	To fully utilize all the functionality of a system, it must be fully populated with data. Many functions of typical systems are dependent on data residing in specific areas or multiple areas of a system; to achieve maximum functionality all data must be populated.
Contain validated data	The initial thought by most organizations replacing a system is to migrate all data from the legacy system with no validation effort. While computer technology has rapidly changed one saying still holds true today: "garbage in, garbage out." All data must be validated for accuracy and applicability. Serious consideration should be given to any data migration effort and cost benefit analysis should be performed.
A Standard Operating Procedure (SOP)	This document captures all decisions, methodologies and processes used to implement, populate and utilize the CMMS in a standardized fashion. This becomes the backbone of the Maintenance Department's policy and procedures manual, as well as, the foundation for the ongoing CMMS training for the site/facility.
An established hierarchy	The hierarchy must be carefully structured to enable cost roll-up. The hierarchy is critical to the CMMS, for it is the foundation by which all report

Table 3-3
Continued

	information is derived. The codes typically established are used as the enablers to sort, search and report information contained in the hierarchy.
Integrated into daily usage	To ensure information is captured, the CMMS must be integrated into the day-to-day activities. All work on equipment, all parts utilized in repair of equipment and all labor costs to accomplish the repair must be captured. The work order is the communication tool the CMMS uses to collect this data. In other words, if the event is not captured in a work order, it never occurred.
System security	Specific system security/access controls must be identified and established. The CMMS informational needs of each role in the organization must be understood and established to ensure the appropriate user has access to all required information while maintaining data integrity.
Cost measures	In order to obtain maximum information, the CMMS must be populated with the burdened labor rate for each individual. This is normally a concern for organizations divulging such information. However, if system security is properly established this is not a concern. The other half of cost control is the materials management function. All materials and purchase costs must be captured. Integration of materials and purchases utilized is essential to enable true cost analysis at the equipment level.

If the system is not going to be populated with labor costs, 50% of the capabilities have been discarded. If the materials management function is not going to be utilized or interfaced, the other 50% of the capabilities have been discarded.

Serious consideration should be given to where implementation assistance is sought. If the issues are software related the software vendor is a likely choice. However, if the issues are maintenance process issues, the software vendor may not be the best choice. The software vendor is a logical choice to assist and train on the functionality of their respective systems. However, they typically lack maintenance experience and tend to approach issues from the software support point of view. Numerous maintenance-engineering firms specialize in the implementation of CMMS and are not software specific in their solutions.

Any system is only as good as the data that it contains, as well as the established flow to ensure daily events are captured. Any system development must include effective consultation with all users prior to implementation, so that system capabilities align well with user responsibilities. All of the following elements must be covered in detail to achieve a singular, effective and integrated CMMS:

- Work order control
- Planning and work measurement
- Materials support
- Preventive maintenance scheduling and leveling
- Scheduling and work assignment
- Equipment history and maintenance engineering support
- Cost accounting
- Budgetary control
- Equipment Data (Equipment Inventory Listing, Nameplate Data, Install Date, see Chapter 3, Section 3.1.6)
- Maintenance Procedural Documentation (written step-by-step work instructions)
  - Equipment Maintenance Plans
  - Preventive Maintenance Procedures
  - Corrective Maintenance Procedures
  - Predictive/Condition Monitoring/Condition (Performance) Testing Procedures
- Output Reporting Formats (Management Reports) and Data (Performance Measures) (see Section 3.1.6 for descriptions and documentation examples)

Each of these categories can have as large, or even larger, listings of individual data fields as the Equipment Inventory Listing. If you haven't planned for implementation or contracted out for it, your chances of actually completing CMMS implementation within the next decade are not high. Because the effort of implementation is so large and because proper implementation of CMMS is so important to the success of a Total Productive Maintenance Program, it is highly recommended that implementation be outsourced to a contractor with significant and successful implementation experience. Ask for references!

Most CMMS users track some level of their maintenance and repair work, but very few track 100% of it. If you are to obtain complete and accurate labor and material cost information and accumulate preventive, corrective and predictive maintenance and failure information for the development of optimized maintenance activity, tracking anything less than 100% of your maintenance and repair work and 100% of maintenance and repair spares is unacceptable.

The emergence of Enterprise Systems (ES)—software packages with fully integrated modules for all the major processes in the entire organization—offers the promise to integrate all the information flows in the organization with the following benefits:

- Replacing a large number of the legacy systems with a single integrated system produces significant cost savings. It eliminates the expensive tasks of maintaining redundant data, transferring data between less than compatible systems and updating and debugging obsolete software code.
- Managers can make informed decisions when data on multiple aspects of operations are readily available for analysis. If the financial-reporting system cannot talk with the maintenance management system, then optimal decisions on equipment replacement cannot be made with confidence. If the work-order control system is incompatible with the inventory control and purchasing systems, then maintenance jobs cannot be done efficiently when the critical spares are not available. Fragmentation of information is a cause of unsupported decisions.

The decision to install a generic, off-the-shelf ES has its pitfalls. Managers must consider the implications on their business imperatives. They should check whether the logic of the system is in conflict with the logic of the organization's practices. The suitability of an ES should be determined from a strategic perspective. In other words, the enterprise should be stressed, not the system. If maintenance is a significant function in the organization, the ES should have modules supporting maintenance management. The required features in these modules include facilities for maintaining records of equipment history, support for preventive maintenance, work-order control, inventory control and purchasing. Through integration with the other software modules that handle payroll, accounts payable, cost accounting, shop-floor data collection, knowledge-base diagnostics, etc., real-time decision-support information can be retrieved by managers using user friendly interfaces.

To leverage the benefits of Enterprise Systems that support maintenance, managers are advised to specify the following requirements in the software modules:

- To exploit the wealth of information embedded in their maintenance data, there should be functions that support modeling of lifetime distributions, inspection or preventive maintenance schedules or equipment replacement decisions. Without these decision supporting attributes, organizations are likely to be data rich, but information poor.
- If RCM (see Chapter 3, Section 3.2) is implemented, features that support the methodology are desirable. Support for documentation of failure modes and effects analysis (FMEA) is one such feature.
- The system must be able to present performance results in a format specified by the user. If the Balanced Scorecard approach (see Chapter 4, Section 2.4.2) is in use, the system should be able to support it. In such cases, the design should follow the logic of the process—strategic objectives are linked to their performance measures, which in turn, have their respective targets; the top-level BSC is deployed to lower level ones in a cascading manner. Navigating within the process should be done through a graphical user interface (GUI). The system should allow the user to drill down high-level measures to reveal further details provided by the lower-level measures they summarize. Trending of data is another essential capability required. Additionally, the information should be accessible in real time to all employees who play a direct role in affecting the performance tracked.
- If the organization has strategic partners in its logistics system, there are huge benefits in establishing direct electronic links with their software systems. If the inventory control, purchasing and accounts-payable modules can communicate seamlessly with their counterparts in your suppliers, then provisioning of spares can be managed efficiently with minimal human intervention and transactions can be processed with low error rates. If part of the maintenance service is outsourced, a direct link with the external supplier's system will shorten the elapsed time between the issue of job requests and response of the supplier. Tapping into the supplier's system also enables the user to monitor the supplier's performance in delivering the required maintenance services. This requirement suggests that the strategic partners need to be involved in establishing the system specifications and in system commissioning.

3.1.7 Maintenance Documentation

Technical Documentation—After creating an up-to-date equipment inventory, the equipment technical documentation should be verified as on-hand and up-to-date. Technical manuals, tech and maintenance

EQUIPMENT MAINTENANCE PLAN ABC MANUFACTURING COMPANY Reading, Pennsylvania Plant #1							CMPRS 1, 2, 7, 8
WORTHINGTON AIR COMPRESSOR MODEL WO-1200							
Maintenance Requirement	Frequency	Procedure	Reference	Craft/Skill	Time	Condition	Related MR
Daily Inspection	D	Daily Walk Thru	None	1 - Mech I	< .5 Hr.	Oper.	None
Weekly Inspection	W	Weekly Walk Thru	None	1 - Mech II	< .5 Hr.	Oper.	None
Obtain Oil Sample	M	M1-CMPRS-WO1200	None	1 - Mech II	.5 Hr.	Oper.	None
Test Cntrlr, Reliefs, Shutdowns	M	M2-CMPRS-WO1200	WO-1200, pp.21-23	1 - Mech I/1 - Elec II	1 Hr.	Shut-down	None
Change Compressor Oil, Check Belt Tension	Q	Q1-CMPRS-WO1200	None	1 - Mech I/1 - Mech II	1 Hr.	Shut-down	M2-CMPRS-WO1200
Etc. ↓							

Figure 3-8 Equipment Maintenance Plan

bulletins, modification and alteration documentation, P & IDs, O & M Manuals and any other applicable documentation should be checked against the equipment inventory list for correct make, model and size, as well as for completed installation of modifications or alterations. The technical documentation is usually the source of repair parts and materials information. If it does not match the as installed configuration, there could be no parts support for that piece of equipment.

Equipment Maintenance Plan—The first operating element to be developed for your TPM Program is the Equipment Maintenance Plan (EMP). These documents define all of the maintenance requirements for each piece of equipment in your plant. There will be one EMP for each unique piece of equipment. The minimum information to be included on each EMP is illustrated in Figure 3-8. The "Related MR" (Maintenance Requirement) column is provided for scheduling purposes MRs that can be performed in conjunction with the listed MR are shown in this column.

Maintenance Task Procedural Documentation—Maintenance procedures provide the systematic guidance for performing each maintenance requirement. Several information sources should be utilized for developing the actual maintenance procedure content. Each source is validated as applicable through comparison with the Equipment Inventory data previously described.

Maintenance Procedure Information Sources:

• Manufacturer's O & M/Technical Manuals
• Existing Maintenance Task Guidance/Documentation
• Manufacturer's Maintenance Bulletins
• ASME, ANSI, AIEE, OSHA and other standards/specifications as applicable

Formatting of procedures is dependent on the type of procedure. There are four general types or categories of Planned Maintenance activities as shown in Figure 3-9.

PREVENTIVE (PLANNED/PROACTIVE) MAINTENANCE TYPES DEFINED	
Preventive	Cleaning, Change oil, Clean/replace filter, Lubricate bearing, etc.
Corrective	Replace bearing, Align coupling, Teardown/overhaul, etc.
Condition/Test	Inspection, Measure alignment, Measure pump head/flow, etc.
Predictive	Oil analysis, Vibration monitoring, Infrared imaging, etc.

Figure 3-9 Planned Maintenance Activities

The first three types, Preventive, Corrective and some Condition/Test Procedures, should contain the following minimum information:

Procedure Number—The number is assigned for use in identifying the procedure and for indexing the procedure when incorporated into the site-specific CMMS.

System Description—This section is text that describes the machine/equipment application.

Procedure Description—Text that describes the procedure purpose. The body of the procedure is often divided into subsections. Each subsection has a heading and that heading is duplicated in the Procedure Description. This ensures that the entire scope of work is well understood.

Related Tasks—Identifies other tasks that should be performed. Usually tasks with a shorter periodicity. For example, an annual procedure will identify any semiannual, quarterly, or monthly procedures to ensure that all work is done during a single equipment maintenance shutdown.

Periodicity—Describes how often the procedure is scheduled. Codes normally used are:

D = Daily, W = Weekly, M = Monthly, Q = Quarterly, S = Semi-Annually, and A = Annually.

Multiples of the above are sometimes used and are identified by a number followed by a letter. For example, 5A indicates that the procedure is scheduled every 5 years.

Craft & Labor (Hrs)—The craft(s) required to perform the procedure is shown and will be followed by two numbers. First is number of people, second is estimated time for each person. For example: 2-people/2 hrs each. The time estimate is to perform the task. Because each site is different, no time was estimated to get to the job site.

Special Tools—Identifies tools and test equipment that the technician will need at the job site. Common tools are not usually identified.

Materials—All materials that will be needed at the job site are listed in this section.

Reference Data—Identifies information, such as a test procedure, that the technician will need in order to perform the task. This section does not identify reference data that may have been used to develop the procedure. Only that reference data needed to perform the task is listed in this section.

Warnings—A warning is identified in the procedure anytime there is the potential for injury (including toxic release to the environment). This section lists every warning that is part of the procedure. If a warning is applicable many times in the procedure, it is shown each time it applies.

Cautions—Similar to the warning, a caution is identified in the procedure anytime there is the potential for damage to the equipment or damage to collateral equipment.

Notes—The procedure may also contain a note. A note provides relevant information to the person performing the procedure.

Preliminary—The first part of the procedure is identified as the preliminary section. This section includes all steps taken before going to the job site, or if at the job site, before starting work on the specified machine. Although there is no maximum number of preliminary steps, this section is usually less than 10 steps.

Procedure—The start of the procedure is clearly identified by the title "Procedure." The first step in the procedure is labeled "A" and is a phrase that identifies the work to be accomplished. The next step is labeled "A1" and is an action item. Each subsequent action step is numbered in ascending order, "A2, A3 . . ." If the procedure can be broken into discrete sections, there may be a "B," "C," etc.

Inspection/Measurement Data—If data is to be collected, there will usually be a Data section. The procedure will identify the data and direct where it is to be recorded in the Data section or other location. The Data section is always located at the end of the Procedure section.

Actual Time—A space to record the actual time that was required to perform the procedure. This information is used by the Planner and Scheduler to refine the estimated time for more accurate scheduling.

At the plant's option, there may also be a place to record the name of the lead person performing the maintenance action. Maintenance personnel skill levels, management requirements and utilization for training determine the level of detail contained in each Maintenance Requirement procedure. Procedures can be an element of maintenance training by incorporating the maximum level of procedural detail with line entries for:

- Gaining access
- Basic skill steps, such as:
- At motor coupling end and on top of bearing housing, remove protective cap from grease fitting
    Wipe grease fitting to remove dirt, moisture and contamination
    Attach fitting on grease gun flex hose over motor grease fitting
    Apply two strokes of grease to motor grease fitting

A typical Maintenance Requirement Procedure follows in Figure 3-10.

The fourth type, Predictive Maintenance (PdM) Procedures, as well as some forms of Condition Monitoring and Equipment Testing, are dependent on how the procedure is performed. Many plants contract for things like Infrared (Thermographic) Imaging or Vibration Measurement/ Analysis. Still others have automated PdM processes and occasionally have automated test and (condition) monitoring equipment built into the manufacturing process systems. For those plants performing these procedures manually, with in-house personnel, the procedural documentation should contain the bulleted information as shown above. In Table 3-4 below

```
                                          MR # Q1-CMPRS-CAC150
SYSTEM: Widget Housing Fabrication & Assembly, Lines 1 & 2
EQUIPMENT: Control Air Compressors 1, 2, 3 & 4

Maintenance Requirement: Check Belt Sheave Alignment, Drive Belt
Condition and Tension
Condition - Shutdown      Time Required - 0.5 Hr. each CAC

Craft:  1 - Mechanic II

SPECIAL TOOLS/EQUIPMENT:
1.  Straight edge              2.  Ruler

Procedure
Preliminary
1.  De-energize[Check De-energized] power source to equipment in
accordance with lockout/tagout procedures.

A.  Check Belt Sheave Alignment
    1.  Place a straight edge or pull a string across
outside face of both sheave surfaces.

    2.  The straight edge or string should touch both inside
and outside edges of both sheaves, if straight edge
or string does not touch all four of these points,
adjust fan and/or motor sheave until properly
aligned.
```

```
D.  Record any deficiencies on Work Order Maintenance & Repair Section
Report.
CAC # 1 _____      Time Used _____
CAC # 2 _____      Time Used _____
CAC # 3 _____      Time Used _____
CAC # 4 _____      Time Used _____
```

Figure 3-10 Maintenance Requirement Procedure

Table 3-4
PdM and Condition Monitoring

Predictive Technique	Application	Problem Detection
Vibration	Rotating machinery, e.g., pumps, turbines, compressors, internal combustion engines, gear boxes, etc.	Misalignment, imbalance, defective bearings, mechanical looseness, defective rotor blades, oil whirl, broken/worn gear teeth, etc.
Shock Pulse	Rotating machinery	Trends of bearing condition
Fluid Analysis (lubricants, coolants, feedwater, etc.)	Lubrication, cooling, hydraulic power, boiler and similar systems	Bearings & housings—wear particles, viscosity, contaminants, etc.
Infrared Thermographic Imaging	Electrical switchboards & distribution/control equipment, steam systems, bearing end caps, power electronics, anything generating heat	Loose electrical connections, worn contacts, faulty insulation, leaky steam traps, hot running bearings, overheating components
Ultrasonics	High pressure systems: hydraulics, steam, pneumatics, etc.	Leaks—through and by: valves, restrictors, fittings, etc.
Spectographic Analysis	Fluid and Gas Systems: liquid cooled circuit breakers, transformers, etc. Gas systems in firing/finishing operations or combustion sys., etc.	Contaminants, chemical constituents and proportion, hazards
Waveform/signature analysis (time and frequency domains)	Power Factor, power quality, Electronics: rectifiers, inverters, power supplies, regulators, etc).	System load characteristcs/Motor faults, degraded circuits, faulty solid state devices, etc.

Table 3-4
Continued

Condition Monitoring/Testing	Application	Analysis results
Materials testing—NDT (non-destructive test), borescopic inspections, eddy current, etc.	Heart exchangers, piping systems, magnetic components (motor cores, etc.)	Corrosion, erosion (wall thickness), fatigue cracking, delamination, etc.
Insulation Tests—megohmmeter, polarization index, surge comparison, impedence, DC HiPot	Motor and Generator windings, electrical distribution and control equipment	Trends of insulation resistance—cleanliness/aging/insulation breakdown
Protective (overload) relay tests/time-travel test	Circuit breakers, transformers, controllers, other circuit protection	Non-Sequential tripping, loss of overload protection, personnel and equipment safety
Dimensional measurements, play and clearance	Sliding, rotating and reciprocating applications	Wear: excessive and trend towards minimum acceptable
Performance testing	Heat exchangers, pumps, compressors, refrigeration systems, conveyor/feed systems, nearly any process system	Loss of efficiency, reduced capacity, reduced speed, reduced quality (loss of tolerance control, trending of performance).

the more common techniques used in PdM and Condition Monitoring are listed together with their application and the kinds of problems they will detect.

3.1.8 Maintenance Engineering

If your plant does not have a Maintenance Engineering section, one should be established. The functions and responsibilities of new or existing maintenance engineering groups should be reviewed and revised to integrate and enhance the proactive maintenance organization. An alarming statistic indicates that up to 70% of equipment failures are self-induced. Finding the reasons for self-induced failures, and all failures, is a responsibility of maintenance engineering. Reliability engineering is the primary role of a maintenance-engineering group. Their responsibilities in this area should include:

- Evaluating Preventive Maintenance Action Effectiveness
- Developing Predictive Maintenance Techniques/Procedures
- Performing Condition Monitoring/Equipment Testing
- Analyzing PM/PdM/CM-CT Data for Optimizing Maintenance
- Employing Engineering Techniques to Extend Equipment Life, Including:
  - Specifications for new/rebuilt equipment
  - Precision rebuild and installation
  - Failed-part analysis
  - Root-cause failure analysis
  - Reliability engineering
  - Rebuild certification/verification
  - Age exploration
  - Recurrence control
- Performing Continuous Evaluation of Maintenance Skills Training Effectiveness
- Performing Selected Elements of Reliability Centered Maintenance (RCM)

3.2 FINE-TUNING TPM USING RELIABILITY CENTERED MAINTENANCE (RCM)

Optimizing maintenance effectiveness is a major objective of Lean Maintenance. TPM objectives focus on maintaining equipment reliability and effectiveness. So how do we fine-tune TPM in order to optimize maintenance effectiveness?

The Reliability Centered Maintenance (RCM) process was evolved within the civil aviation industry to fulfill this precise need. In fact, the definition of RCM is:

> A process used to determine the maintenance requirements of physical assets in their present operating context.

In essence, we have two objectives: determine the maintenance requirements of the physical assets within their current operating context and ensure that these requirements are met as cheaply and effectively as possible. RCM is better at delivering objective one; TPM focuses on objective two. By incorporating elements of RCM into our TPM Maintenance Operation we can make the necessary refinements to achieve Lean Maintenance.

3.2.1 What RCM Accomplishes

Reliability Centered Maintenance (RCM) is a continuing process used to determine the most effective approach to maintenance in support of the mission. It identifies the optimum mix of applicable and effective maintenance tasks needed to realize the inherent design reliability and safety of systems, equipment and personnel at minimal cost (a goal of TPM). RCM uses a systematic, logic based approach for determining objective evidence for selecting the most appropriate maintenance tasks.

RCM generates sound technical rationale and economic justification on which maintenance decisions are based. The process considers operational experience and failure history to generate, validate and support those decisions.

3.2.1.1 The Origins of RCM

The original development of RCM concepts is generally attributed to maintenance policy events in the airline industry in the late 1960s and early 1970s. In an effort to maximize the safety of airplane passengers and maximize the reliability of aircraft and aircraft equipment, a task group was formed to investigate maintenance practices and to challenge the traditional concepts of successive overhauls. The traditional concept promoted the belief that every item on a piece of complex equipment degrades over time, and that a specified age can be defined where overhauling that equipment would ensure safety and operating reliability. The resultant work of this task group demonstrated that a strong correlation between age and failure rate did not exist and that the basic premise of time-based maintenance was false for the majority of equipment. The results of this task group's investigation can be summarized in the following three significant discoveries:

- Scheduled overhaul had little effect on the overall reliability of equipment unless the item had a dominant failure mode and the maintenance action directly addressed that dominant failure mode.
- There were many items for which no effective form of scheduled maintenance could be identified.
- Cost reductions in maintenance could be achieved without a decrease in reliability. In fact, a better understanding of the failure process of complex equipment would actually improve reliability when unnecessary maintenance actions were eliminated.

The traditional approach to Preventive Maintenance was to overhaul or replace components just before they "wore out" and caused an equipment failure. The trick was to accurately determine where this "wear out" point was in the life cycle of the component. Studies performed in the civil aircraft industry indicate that all components do not follow a "operate reliably then wear out" failure probability. They tend to follow a variety of failure probabilities, as illustrated in Figure 3-11.

The failure probability distribution must be considered when defining a PM/PdM strategy for equipment. While an overhaul or replacement strategy may be appropriate for components that have an age-related nature of failure, it will have no effect on the reliability of components whose failure distribution is according to one of the lower three curves. In fact, if an overhaul is done based on an assumed end-of-life when the component follows the "worst new" distribution, the overhaul itself is likely to cause an "infant mortality" failure.

Although failure probability data is limited outside the aircraft industry, the available information suggests that the more complex the equipment, the more it behaves like the bottom two curves. This would imply that a predictive inspection strategy would be most appropriate for most of today's complex manufacturing systems.

3.2.1.2 Properties of RCM:

RCM is not a new strategy by which organizations embrace maintenance, but rather it is a combination of three distinct approaches to maintenance. Those approaches are

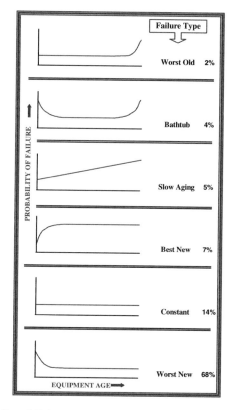

Figure 3-11 Probability of Failure

- Reactive Maintenance, which consists of repair actions after failure,
- Planned Preventive Maintenance (PM) strategy, or "time-directed" maintenance, which consists primarily of health maintenance actions prior to failure
- Predictive Maintenance (PdM) and Condition-Based Maintenance (CBM), which consists of measurement and monitoring of equipment conditions to facilitate prediction of equipment/component failure

The RCM process entails asking seven questions about the asset or system under review, as follows:

1. What are the functions and associated performance standards of the asset in its present operating context?
2. In what ways does it fail to fulfill its functions?
3. What causes each functional failure?
4. What happens when each failure occurs?

5. In what way does each failure matter?
6. What can be done to predict or prevent each failure?
7. What should be done if a suitable proactive task cannot be found?

The Primary Principles of RCM are:

RCM is Function Oriented—It seeks to preserve system or equipment function, not just operability for operability's sake. Redundancy of function, through multiple equipment, improves functional reliability, but increases life cycle cost in terms of procurement and operating costs.

RCM is System Focused—It is more concerned with maintaining system function than individual component function.

RCM is Reliability Centered—It treats failure statistics in an actuarial manner. The relationship between operating age and the failures experienced is important. RCM is not overly concerned with simple failure rate; it seeks to know the conditional probability of failure at specific ages (the probability that failure will occur in each given operating age bracket).

RCM Acknowledges Design Limitations—Its objective is to maintain the inherent reliability of the equipment design, recognizing that changes in inherent reliability are the province of design rather than maintenance. Maintenance can, at best, only achieve and maintain the level of reliability for equipment, which is provided for by design. However, RCM recognizes that maintenance feedback can improve on the original design. In addition, RCM recognizes that a difference often exists between the perceived design life and the intrinsic or actual design life, and addresses this through the Age Exploration (AE) process.

RCM is Driven by Safety and Economics—Safety must be ensured at any cost; thereafter, cost effectiveness becomes the criterion.

RCM Defines Failure as Any Unsatisfactory Condition—Therefore, failure may be either a loss of function (operation ceases) or a loss of acceptable quality (operation continues).

RCM Uses a Logic Tree to Screen Maintenance Tasks—This provides a consistent approach to the maintenance of all kinds of equipment.

RCM Tasks Must Be Applicable—The tasks must address the failure mode and consider the failure mode characteristics.

RCM Tasks Must Be Effective—The tasks must reduce the probability of failure and be cost effective.

3.2.2 Integrating RCM and TPM

Fine-tuning Total Productive Maintenance through integration with elements of Reliability Centered Maintenance does entail doing some things a little differently in our maintenance operation. The differences are what characterize our maintenance operation as Lean. The first two actions toward integration will require:

- Assessing Equipment Criticality
- Establishing Maintenance Task Priority Codes

3.2.2.1 Equipment Criticality and Maintenance Priorities

Criticality Assessment provides the means for quantifying how important an equipment or system function is relative to the identified mission—normally production, although environmental and safety systems also require assessment. Table 3-5 provides one method for criticality ranking providing 10 categories of criticality—or severity. It is not the only method available. The categories can be expanded or contracted to produce a site-specific listing.

A worksheet method of assigning point values to various criticality categories and ranking equipment and systems by total points is another method that is often used to assess equipment criticality. A sample worksheet illustrating this method is provided in Appendix B.

Maintenance Task Prioritization for assigning Work Order Priority is the second step of integrating RCM and TPM. The number of Work Order priorities and their characteristics, while somewhat discretionary, are normally arranged as shown in Table 3-6 below.

Planning and Scheduling—When planning maintenance work, the equipment with the highest criticality is scheduled first and maintenance tasks are scheduled in order of priority until all available maintenance resources have been utilized. When PM/PdMs that are due to be performed are consistently deferred because of a lack of resources, augmentation of the maintenance work force is indicated.

3.2.2.2 Reliability Engineering

In fine-tuned TPM, or Lean Maintenance operations, the Maintenance Engineering Group takes on the additional responsibility of performing Reliability Engineering. In combination with other proactive techniques, reliability engineering involves the redesign, modification or improvement of components or their replacement by superior components. Sometimes a complete redesign of the component is required. In other cases, actions such as upgrading the type of component metal, adding a sealant or some similar fix is all that is required. Progressive maintenance engineering groups include a specifically trained reliability engineer assigned

Table 3-5
Criticality Assessment

Description of Failure Effect	Effect	Ranking
No reason to expect failure to have any effect on Safety, Health, Environment or Mission.	None	1
Minor disruption of production. Repair of failure can be accomplished during trouble call.	Very Low	2
Minor disruption of production. Repair of failure may be longer than trouble call but does not delay Mission.	Low	3
Moderate disruption of production. Some portion of the production process may be delayed.	Low to Moderate	4
Moderate disruption of production. The production process will be delayed.	Moderate	5
Moderate disruption of production. Some portion of production function is lost. Moderate delay in restoring function.	Moderate to High	6
High disruption of production. Some portion of production function is lost. Significant delay in restoring function.	High	7
High disruption of production. All of production function is lost. Significant delay in restoring function.	Very High	8
Potential Safety, Health or Environmental issue. Failure will occur with warning.	Hazard	9
Potential Safety, Health or Environmental issue. Failure will occur without warning.	Hazard	10

this responsibility on either a full- or part-time basis, depending on the size of the plant.

Rigorous RCM Analysis has been used extensively by the aircraft, space, defense and nuclear industries where functional failures have the potential to result in large losses of life, national security implications and/or extreme environmental impact. A rigorous RCM analysis is based on a detailed Failure Modes and Effects Analysis (FMEA) and includes probabilities of failure and system reliability calculations. The analysis is

Table 3-6
Arrangement of Work Order Priorities

Priority Code	Classification/Priority/Condition/Description
E Emergency:	*Must be performed immediately.* Higher priority than scheduled work, critical machinery down or in danger of going down until requested work complete. "E" to be used only if production loss, delivery performance, personnel safety (new and imminent), equipment damage or material loss are involved and no bypass is available *Start immediately and work expeditiously and continuously to completion, including the use of overtime without specific further approval.* Only personnel authorized to approve overtime can assign "E" priority to work orders. Emergency work order reports will be sent to plant manager for review.
1 Urgent:	*Needed within a few hours, by end of shift at latest.* In judgment of authorizer, work must be completed as soon as possible but does not require immediate attention. Mechanic(s) assigned as soon as available without halting a job already in progress. Overtime approval is not implied by "1" priority. Overtime authorization must be obtained by special request, given specific circumstances, from the authorizer or designated authority. Priority "1" should be used for equipment which is down or in danger of going down and which affects ability to produce desired product mix or renders plant void of backup capacity in event of subsequent failure.
2 Critical:	*Needed within 24 hours.* Similar to urgent (1) jobs but with less urgency. Typically good work to leave for off-shift coverage personnel. Controlled use of "E," "1" and "2." Priorities must be reserved for truly critical situations or they diminish planning and scheduling effectiveness.
3 Rush:	*Must be performed before end of current week.* Normally, this work will be scheduled to start within 24 to 48 hours after receipt of the work order. Priority "3" jobs (as well as "E," "1" and "2" jobs) cannot be effectively planned before scheduling. All jobs should be assigned priority "4" or higher whenever possible. Priority "3" jobs will be used as fill-in work for personnel responding to emergency and

Table 3-6
Continued

Priority Code	Classification/Priority/Condition/Description
	urgent work orders or will be forced into the current week's schedule, "bumping" a properly planned job already on the schedule. Performance on the job will be measurably less and cost measurably more than if planned (Priority "4" or "5").
4 Essential	*Deferrable—must be performed before the end of next week.* This work can be effectively planned. It will be scheduled next week (as opposed to being scheduled in the order of request date). Realize that all requests cannot be completed next week. Priority "4" requests delay the completion of previously requested work of lower priority and drive up the cost of requested work as overtime will be requested to meet the time/demand constraints implied by priority. Use Priority "5" if possible.
5 Desirable:	*Designates desirable but deferrable jobs. Can be completed anytime within the next few weeks,* and can therefore be scheduled on the basis of first-requested/first-scheduled. A desired completion date may be indicated by the originator. Weeks of backlog report provide the current wait to be anticipated on Priority "5" jobs. Overtime will not be used on Priority "5" jobs unless the work must be performed on a non-operating day or aging of the request exceeds six weeks.
6 Shutdown:	A "6" priority designates work requiring programmed shutdown. Work orders in this category are accumulated for shutdown planning.
7 Routine:	Used exclusively for routine work—usually on standing work orders, "7" is not associated with normal day-to-day work order requests.

used to determine appropriate maintenance tasks to address each of the identified failure modes and their consequences. The considerations of a rigorous RCM analysis are illustrated in Figure 3-12.

While this process is appropriate for these industries, it is not necessarily the most practical or best approach to use for manufacturing

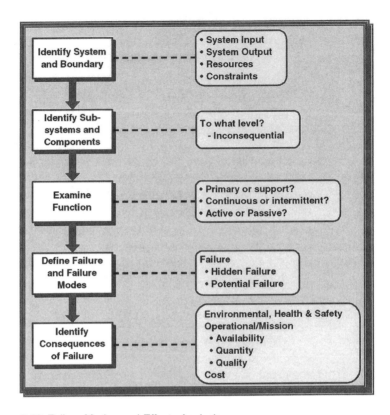

Figure 3-12 Failure Modes and Effects Analysis

systems maintenance. For these systems a streamlined or intuitive RCM analysis process may be more appropriate. This is due to the high analysis cost of the rigorous approach, the relative low impact of failure of most facilities systems, the type of systems and components maintained and the amount of redundant systems in place. The streamlined approach uses the same principles as the rigorous, but recognizes that not all failure modes will be analyzed.

Nonetheless, because your plant may have equipment or production systems that qualify for Rigorous RCM Analysis, Appendix A contains a description of the FMEA process and related calculations along with a sample FMEA Data Sheet for recording the results.

The Probability of Occurrence (of Failure) is based on work in the automotive industry. Table 3-7 provides one possible method of quantifying the probability of failure.

Table 3-7
Probability of Occurrence (of failure)

Ranking	Effect	Comment
1	1/10,000	Remote probability of occurrence; unreasonable to expect failure to occur
2	1/5,000	Low failure rate; similar to past design that has, in the past, had low failure rates for given volume or load
3	1/2000	Low failure rate; similar to past design that has, in the past, had low failure rates for given volume or load
4	1/1000	Occasional failure rate; similar to past design that has, in the past, had similar failure rates for given volume or load
5	1/500	Moderate failure rate; similar to past design that has, in the past, had moderate failure rates for given volume or load
6	1/200	Moderate to high failure rate; similar to past design that has, in the past, had moderate failure rates for given volume or load
7	1/100	High failure rate; similar to past design that has, in the past, had high failure rates that have caused problems
8	1/50	High failure rate; similar to past design that has, in the past, had high failure rates that have caused problems
9	1/20	Very High failure rate; almost certain to cause problems
10	1/10+	Very High failure rate; almost certain to cause problems

If there is historical data available, it will provide a powerful tool in establishing the ranking. If the historical data are not available, a ranking may be estimated based on experience with similar systems in the facilities area. The statistical column (Effect) in the table can be based on operating hours, day, cycles or other units that provides a consistent measurement approach. Likewise, the statistical bases may be adjusted to account for local conditions. For example, one organization changed the statistical approach for ranking 1 through 5 to better reflect the number of cycles of the system being analyzed.

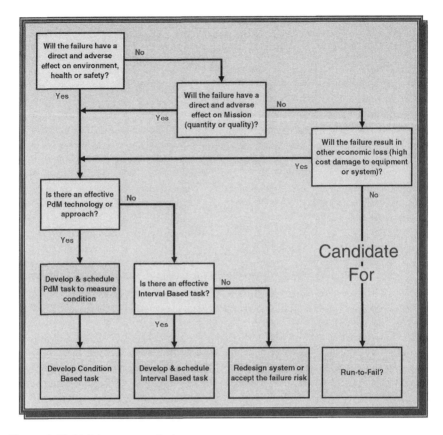

Figure 3-13 Maintenance analysis process

RCM acknowledges three types of maintenance tasks. These tasks are:

- time directed (PM)
- condition-directed (PdM)
- failure finding

Time-directed tasks are scheduled when appropriate. Condition-directed tasks are performed when conditions indicate they are needed. Failure finding tasks detect hidden functions that have failed without giving evidence of pending failure.

Run-to-Failure is a conscious decision and is acceptable for some equipment.

Note that the maintenance analysis process, as illustrated in Figure 3-13, has only four possible outcomes:

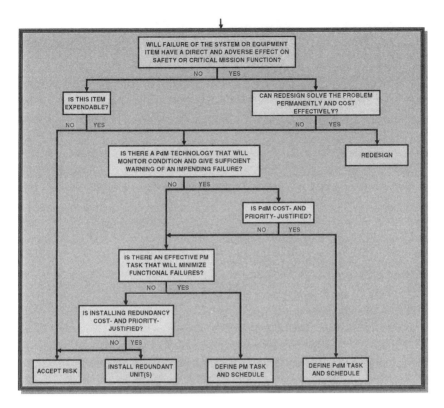

Figure 3-14 Abbreviated decision tree used to identify the maintenance approach

- Perform Interval (Time- or Cycle-) Based actions
- Perform Condition-Based actions
- Perform no action and choose to repair following failure
- Determine that no maintenance action will reduce the probability of failure AND that failure is not the chosen outcome (Redesign or Redundancy)

In a formal RCM operation, analysis of each system, subsystem and component is normally performed for all new, unique and/or high-cost systems. This approach is not utilized in fine-tuned TPM (Lean Maintenance). Instead, an abbreviated decision tree, such as the one illustrated in Figure 3-14 is used to identify the maintenance approach.

Regardless of the technique used to determine the maintenance approach, the approach must be reassessed and validated. The iterative RCM decision process can be used for a majority of manufacturing plant systems and equipment.

At the system level, the determination of whether there is suitable PdM technology available to provide warning of impending failure requires the reliability engineer to break the system down to the component level and further analyze functional failures.

4

Pre-Planning for Lean Maintenance

4.1 GAINING KNOWLEDGE/IMPARTING KNOWLEDGE

It should be obvious that before you undertake transformation to a Lean Maintenance operation you should be more than "just familiar with the general concepts." In order to acquire in-depth knowledge in the implementation of Lean Practices you will need more than attendance at a one-week seminar on the principles of Lean. Ideally, the knowledge of lean implementation would be the result of having been involved in a successful implementation. However, it is unlikely that many manufacturing plant maintenance organizations will have an asset with that kind of experience. But that level of know-how is critical to the success of your transformation. The alternative is to bring on a consultant with that kind of first-hand knowledge or to know and understand the contents of this handbook backwards and forwards and believe in it.

4.1.1 Selecting the Lean Maintenance Project Manager

The next step in the Lean Maintenance transformation is the assignment of the plant's project manager (PM) for the transformation. This is not a part-time job and the PM will not be able to split his time between Lean Maintenance PM functions and another assignment within the plant. He must be a dedicated asset to the Lean transformation. There are several critical personal traits and characteristics that the selected PM should possess if he is to be successful in leading your maintenance organization into the era of Lean.

First and foremost, the PM must have the authority, originating from top management, to completely and effectively perform as the Leader for the Lean Maintenance Transformation.

4.1.1.1 Necessary Attributes of Lean Maintenance PM

He must also be:

- Known/Respected by management and by work force
- Knowledgeable in
  - Maintenance (TPM) Process
  - CMMS Operation
  - Planning and Scheduling
  - MRO Storeroom Processes
- Tough but fair
- Self-Motivated and a Visionary Type of Person
- Insistent on Continuous Improvement
- Able to Interact Effectively with all Departments and all Levels
- Ultimately the Lean "Knowledge Expert"
- Constantly learning more about Lean
- Constantly Communicating and Promoting Lean
- A True Believer in Lean Processes

4.1.1.2 Lean PM Duties and Responsibilities

As the leader of the transformation process to Lean Maintenance, the Lean PM has a multitude of duties and responsibilities. Some are to be exercised directly and others, by necessity must be applied through various managers. For this reason it is important that the plant's upper management thoroughly indoctrinate departmental managers in the role of the Lean PM and in the need for a cooperative and committed relationship to the Lean PM and his responsibilities. This in no way should be seen as usurping the manager's authority, because he still retains his own responsibilities.

- Directly Exercised Responsibilities
- Lead the Transformation via Roadmap
- Educate, Motivate and Direct Transformation (Project) Team Leaders
- Provide Periodic Progress Reports to Management
- Interface in Interdepartmental Issues

- Constantly Communicate and Publicize Lean
- Responsibilities Exercised Through Liaison with Department Managers
- Organizational and Related Changes Needed to Support Transformation
- Fundamental Process Changes Required (e.g., Work Order System)

4.1.2 What You (the Lean PM) Should Know

Of course you will need to be familiar with all the principles and processes of Lean thinking and Lean implementation. If you are working with an expert consultant, one who has had experience in a successful Lean transformation, your role initially will be to translate the practices of Lean manufacturing into maintenance practices. At the same time you will be undergoing a transformation of your own—into a Lean Maintenance Expert. The Lean principles and tools discussed briefly in this section will be described in detail in Chapter 6—Mobilizing and Expanding the Lean Transformation.

a. To begin, you will need to have in-depth knowledge of the five steps or principles of Lean Implementation.

Step 1: Specify Value

Define value from the perspective of your customer as well as the product or final customer. Your internal customer is production and production equipment operators. Your external customer is the product consumer. Express value in terms of a specific product, which meets the customer's needs at a predefined cost and at a specific time.

Step 2: Map

Identify the value stream, the set of all specific actions required to bring a specific product through the three critical management tasks of any business: the problem-solving task, the information management task, and the physical transformation task. Create a map of the Current State and the Future State of the value stream. Identify and categorize non-value-adding waste in the Current State, and eliminate it!

Step 3: Flow

Make the remaining steps in the value stream flow. Eliminate functional barriers, interruptions, detours and back-flows and develop a product-focused organization that dramatically improves lead-time. Maintenance must be ready to proceed at scheduled equipment availability time and equipment restored to production (in spec) at scheduled on-line time.

Step 4: Pull
Let the customer pull products as needed, eliminating the need for a sales forecast. In maintenance, perform maintenance on on-line production equipment not on off-line equipment; maintenance tasks that sustain production tolerance/quality specifications are a priority.
Step 5: Perfection
There is no end to the process of reducing effort, time, space, cost and mistakes. Maintenance tasks are performed correctly the first time and every time. Maintenance task completion restores equipment to production specifications and tolerances (a good example of Maintenance/Production cooperative relationship). Return to the first step and begin the next Lean transformation, offering a product, which is ever more nearly what the customer wants.
b. The PM for the Lean Maintenance Transformation should also have a good general knowledge of the tools available for the Lean transformation.
5-S Process
Seiri—Sort what is not needed. Use a color-coded tagging system; red tags for items considered not needed. Then provide everyone a chance to indicate if the items really are needed. Red tagged items that no one has identified a need for is eliminated.
Seiton—Straighten what must be kept. Make things visible, e.g., put tools on pegboard and outline the tool's shape so its storage location can be readily identified. Apply "a place for everything and everything in its place" philosophy.
Seiso—Scrub everything that remains. Clean and paint to provide a more tidy and pleasing appearance.
Seiketsu—Spread the clean/check routine. When others see the improvements, provide with training and the time to improve their work area.
Shitsuke—Standardization and self-discipline. Develop a cleaning schedule. Use downtime to clean and straighten area.
Seven Deadly Wastes
1. Overproduction—Doing more PM than is needed, gaining little or no demonstrable reliability improvement. Are crews maintaining equipment that should instead be allowed to run-to-failure and then be replaced? We often find maintenance mechanics going out and re-checking equipment over and over without ever finding either any deterioration or failure—and yet, the PM's haven't changed in years. That's unnecessary "overproduction." Pointless effort of this kind can work in either direc-

tion, i.e., PM is still done despite recurring breakdowns—which means that the PM was useless all along—or PMs are being done "ritually" for systems that rarely fail, whether inspected or not, or which fail at predictable intervals. In either case, PMs are having little or no effect. PM only "adds value" if it enhances equipment longevity. You have to constantly reevaluate and reassess these relationships. Look at the failure histories that drive the PM schedule. Be sure that they're still reasonable and appropriate. Other examples of "overproduction" include items such as excessive recordkeeping, tracking, looking or assorted busy-work.

2. Waiting—When maintenance personnel are forced to sit idly for parts to come, or wait for some other event. When people aren't in motion and they should be, they're not adding any value. That is waste, and lack of good coordination between task elements is a common cause in maintenance. One solution: Have schedulers get more feedback from the crews to understand where bottle-necks or mistiming happens and why. Waiting is caused in scores of other ways too, such as from outdated or bad work procedures, or lack of training. For example, an operator's machine goes down and needs repair. The operator summons a nearby mechanic, and the two spend half an hour troubleshooting in futility, because the operator didn't know (or care) that the proper procedure is to report the problem to a maintenance scheduler, or the operator didn't know how to create an accurate work order and enter it on the computer terminal. To eliminate these waste areas, you have to explore all the steps that occur, and then identify those that add value or contribute to progress in responding efficiently. Look at what is necessary, and look at what can be eliminated.

3. Transportation—Ask anyone in the plant what he sees mainte-nance people doing and he will often answer walking or driving around. Tools stored far from the job or task-at-hand, common or repetitive use of parts that have not been preassembled or kitted, documentation that must be found and work orders for machines that are not available for shutdown are common causes. Each activity requires transportation and most do not add value to the maintenance process. Crisscrossing and long transit times, with techs shuffling back and forth to stores or to engi-neer's offices, etc., or doing tasks with elements unnecessarily spread out, are major sources of waste. In sprawling facilities the cumulative loss can be high.

As for machine documentation, typically, you may want to set up some kind of centralized library system where all documents are well organized. Whether the maintenance person or the Planner obtains documentation, such a system can eliminate much of this excess motion. In addition, make copies of key data sheets and diagrams, etc., to hang next to each pertinent piece of equipment. Efficiency gains from these moves will be compounded by improved response times, faster detection of imminent failure, increased productivity, higher machine reliability and better overall communications.

4. Processing—In this category are the typical maintenance bottle-necks, such as an inefficient or haphazard work order system, excessive or time-consuming reporting forms, ineffective train-ing—which fails to convey needed instructions and must be continually retaught, etc. Studying the problem can develop solutions to wasteful processing. Typical measures may include a CMMS upgrade, workflow process analysis and work flow reassessment. Streamlining, clarity of instructions, better recog-nition of actual root problems, improved technical training, systematizing of work order forms, and better work planning through the use of a dedicated Planner/Scheduler and instilling a better work ethic and the determination to "do things right every time."

5. Inventory—Quite often a major contributor to waste in both time and overhead cost is repair parts and storage. Frequent "stock-outs," large stockpiles of obsolete parts and large inven-tories of infrequently used and/or expensive or limited shelf life items are most common. Formulas are available for determining appropriate stock levels and reorder points. Integration of MRO storeroom inventory control with CMMS is exceptionally effec-tive in eliminating this area of waste. Also consider storeroom layout and processing flow and item cataloging efficiency so that retrieval takes minimal time. Are rooms logically arranged, with high-demand items quick to get? How about shelves and bins? Are they well-marked, so that parts are easily accessible? Good storeroom management brings such impact that bringing in a consultant or obtaining training will easily justify the investment, especially if your system hasn't recently been examined.

6. Motion—Maintenance personnel also often burn inordinate time searching for key information: schematic diagrams, manuals, parts lists, repair histories—all of which are critical to servicing, but scattered and unorganized. To plug up such gaps, do some

basic time-and-motion studies, quantifying the unnecessary movement factors. Give thought to how your parts, stores, tools, people and equipment might be better repositioned to make them closer, handier and logistically more accessible. For example, in Lean Manufacturing a common "proximity improvement" is called point-of-use tooling, i.e., "getting tooling and supplies close to where they are being used, as opposed to having them off somewhere else thoughtlessly.

7. Defects—Relative to maintenance defects are instances of reworking, redoing and repeatedly repairing an item due to failure to identify root cause of a failure. Defects in the maintenance operation also revolve around preventive maintenance tasks that do not add value to the output. For example, a quarterly oil change on a machine that has not been operated in three years should be extended based on actual lubricant condition as determined by oil analysis. As in processing, defects like these are solved by studying the problem. Creation of a maintenance engineering function whose responsibilities include root cause failure analysis and maintenance effectiveness evaluations to identify ineffective or incorrect maintenance task procedures or incorrect scheduling (frequencies) can quickly identify and correct this type of waste.

Standardized Work Flow—The standard is the best, easiest and safest process to complete the job. Components include:

TAKT (Cycle) Time (TAKT is German for pace)—In Production it is the available work minutes divided by required product quantity to determine minutes per piece or cycle time. In Maintenance it is the available work minutes for scheduled maintenance divided by the scheduled time required.

Work Sequence—Individual steps in the process being performed by employees

Standard WIP (Work in Process)—Smallest amount of WIP required to do the job

Value Stream Mapping—A process mapping technique. By mapping a complete process and then mapping just the value-added steps, we gain insight into the amount of waste inherent to the process. Approximations for North American manufacturing industry production processes are:

5% value adding activity

60% non-value adding

35% unavoidable non-value adding

Just-in-Time (JIT) and Kanban (Pull System/Visual Cues)—In the manufacturing process, one phase of production can "push" his finished subassembly to the next phase, or the next phase can "pull" from the previous phase, as subassemblies are needed. Kanban is the "pull" operation.

Kanban scheduling systems operate like supermarkets—A small stock of every item sits in a dedicated location with a fixed space allocation. Customers come to the store. They visually select and purchase their items. At the checkout station an electronic signal goes to the supermarket's regional warehouse which details which items have sold. The warehouse prepares a (usually) daily replenishment delivery of the exact items sold.

In manufacturing processes, the usage signal can be via computer systems or by visual cues. For example an empty input bin is a cue to fill the bin to a predetermined level or actual visual cue cards can be used to signal the need to replenish input items.

Jidoka (Quality at the Source)—Quality is always a very complex concept and it applies not only to our final product, but also to each condition, operation and action. Traditionally, inspection was the way to prevent a defective piece from leaving the plant. Today "jidoka" is the real concept of quality.

Jidoka means "Autonomous Control" and that is what "Quality at the source" is about. We achieve Jidoka when we create ideas that will help to stop the production when a tool or material is out of spec, when the pressure is not enough to stamp a part with total quality. Quality control applied before we have a bad product.

Poka Yoke is Japanese for Mistake-Proofing. Prevention stops losses before they occur.

Shewhart Cycle. A fundamental element of Lean Thinking is continuous improvement, particularly as applied to people, products, processes, services and all related learning. A quality tool that embodies this concept is the Shewhart Cycle of Learning and Improvement, commonly known as the PDSA or PDCA cycle. You may often see this referred to as the Deming Cycle. Walter Shewhart was a teacher and mentor of Dr. W. Edwards Deming.

PDSA or PDCA stands for Plan—Do—Study (or Check)—Act. This is a way to achieve the outcomes you desire through heightened quality awareness. It is a virtually never-ending process and is key for achieving transformation. The problem solving technique is usually displayed as a continuous circle icon in Figure 4-1.

Figure 4-1 Problem-solving technique

4.1.3 Who Else and How to Familiarize Support Activities

Once the plant's Lean Maintenance Project Manager has been assigned, his work begins virtually immediately. He must develop and deploy the actual "program" or project team—this is a team of the maintenance and maintenance support leaders—and begin their education. Even more importantly he must begin to gain their commitment. Recommended initial assignments are:

- Maintenance Management Representative
- Supervisor From Each Maintenance Work Center
- Manager/Supervisor of Maintenance Storeroom
- IT CMMS Supervisor/Custodian of CMMS
- Production Department Management Representative
- Maintenance Planner and Scheduler
- Maintenance Engineering Leadership

Familiarization and indoctrination of the project team should be conducted as a group. When even one member is absent, the group starts to lose its identity. These are the leaders for the Lean transformation and they can make or break the effort. Without their enthusiastic support and willingness to provide downstream leadership, their team becomes ineffective. For these reasons, careful consideration of their assignment is warranted. If you detect skepticism or indifference in any of your selections, that individual should immediately be replaced.

4.1.3.1 Educating the Project Team

Once the project team membership is finalized, a series of at least three familiarization-training sessions should be conducted with the group. Each session should be scheduled for a minimum of half a day (four hours). Sessions should provide information, as outlined following, in significant detail and as an interactive exchange. All material covered should also be provided in written form; the use of a loose-leaf binder will facilitate retention, note taking and future referral to the material.

Session 1: Purpose of Lean Maintenance Transformation and Preeminent Principles
- Definition (see section 1.4.2) emphasize job security and empowerment
- Top to bottom commitment to their job—to gain commitment
- Role and Responsibilities of Lean Maintenance PM & Project Team
- Waste Elimination and Doing More With Less—7 deadly wastes
- Cooperative Operator/Maintenance Technician Relationships
- Overview of Tools for Lean Maintenance Transformation

Session 2 (two or more sessions may be required for this material): Using Lean Maintenance Implementation Tools
- Process Description of:
  - 5-S Visual
  - Standardized Work Flow
  - Value Stream Mapping
  - JIT and Kanban (Pull) System
  - Jidoka (Quality at the Source)
  - Shewhart Cycle (PDSA)
- Overview of Lean Maintenance Transformation Roadmap

Session 3: Roles and Responsibilities of Lean Transformation Project Team Members
Making Action Team Assignments
Empowering Teams

4.2 THE TRANSFORMATION ROADMAP

Phase 1: Lean Assessment Phase (2 to 4 Months)
The purpose of Phase 1 is to make sure that the Maintenance Department is prepared and ready for Lean Maintenance. It is an evaluation and analysis—then fix operation.
Phase 1 Activities:

1. Evaluate TPM Effectiveness (Project Team to accomplish)
   - Assessment of Equipment Reliability
   - Maintenance Organization Structure
   - Work Order System

- MRO Storeroom Operations
- Planning and Scheduling
- CMMS
- Documentation
- Maintenance Engineering

Your Total Productive Maintenance operation doesn't (not yet) need to be operating in the Maintenance Excellence zone, but it does need to be operating effectively. If you can't find weak areas in your TPM then you may not be doing a critical enough evaluation. One of the primary factors in the make-up of the project team is to have several sets of eyes outside the area being evaluated peering closely at each operation. Knowing what Lean is all about, the team members should especially look for flaws that could impede the implementation of Lean Maintenance.

The evaluation results should be in the form of a brief written report with a laundry list of areas or processes that need to be improved or upgraded together with the criteria for satisfactory progress in order to proceed.

2. Execute Fixes for Each Item on Laundry List
   - Implement Fix (including equipment reliability upgrade)
   - Measure Results
   - Repeat Until Satisfactory
3. Critical Analysis of TPM (as improved) Effectiveness

The critical analysis is indeed critical; the results will determine whether you will proceed with the implementation of Lean Maintenance. If your plant has already committed to the implementation of Lean Manufacturing—and since a Lean Maintenance operation is a prerequisite of Lean Manufacturing—you have no other option (such as abandoning the transformation) than to repeat the evaluate-fix-critically analyze process until your TPM is effective enough to support the Lean transformation.

Phase 2: Lean Preparation Phase (2 to 6 Months)
This is primarily an education phase. The success of this phase is also critical to the journey. As with previous "education" efforts, this phase is also a selling effort. The support of the remainder of the Maintenance Department as well as maintenance support functions depends on the success of Phase 2. Phase 2 is initiated with a kick-off meeting to introduce the Lean Maintenance effort. As far as the rest of the maintenance department and maintenance operation support functions are concerned, this marks the beginning of the Lean Maintenance Transformation.

In preparing for the "kick-off" of the program, the Lean PM and the Project Team must designate initial Action Teams that will perform the first waste elimination exercises or Kaizen Events. A minimum of three teams, one primary and two or more follow-on teams will be required. Membership should be based on the maintenance organization's structure, but should include at least one production line operator, one person from the maintenance storeroom and one person from maintenance engineering, in addition to maintenance techs. These teams will perform the first Kaizen Events for Phase 3.

Phase 2 Activities:

1. Lean Maintenance Kick-off Meeting for the Masses (Orientation)
   • Purpose of Lean Maintenance
   • Principles
   • Action Teams
   • Preliminary Assignments
   • Empowering Teams
2. Lean Leadership Workshops for:
   • Maintenance Manager (Leader and Participant)
   • Production Manager
   • Purchasing Manager
   • IT Manager
3. Lean Principles and Techniques Workshops
   • Action Teams
4. On-site Visitations to Companies with Lean Maintenance
5. Opportunities for One-on-One Sessions for Reassurance

Phase 3 Pilot Phase (1 to 3 Months)

After general indoctrination, Leadership education and assignment and preliminary training of Action Teams is the first test of Lean Implementation. An initial, or pilot Kaizen Event of five to ten days is used to kick-off the Lean Maintenance Transformation. The first Kaizen Event will generate much interest and therefore should serve to reinforce the overall concept of Lean Maintenance. The selection and execution of the Kaizen Event needs to be openly and publicly shared. One Action Team will execute, but all will share in the lessons learned in order to apply them in their own Kaizen Event.

A fundamental element of the Lean structure is Empowering Teams. Action Teams are Empowering Teams. The word "empowering" is used as both an adjective and a verb in the context of Empowering Teams. As an adjective it describes a kind of team. An empowered team is one that plans, carries out and improves their value-adding processes. As a verb it says that you have empowered the team with authority, accountability, direction determination and self-determination (empowered to succeed).

Phase 3 Activities:

1. Pilot (Initial) Kaizen Events

Each Action Team should perform a pilot or initial Kaizen Event during Phase 3. The Primary Action Team executes their event as a solo, which is followed by the Follow-on Action Team Events, which are performed simultaneously. Through the publicly shared reviews of these pilot events, (where everyone is likely to have and offer advice) invaluable training is realized. The pilot events also serve to build confidence and at the same time illustrate that there is much to learn along the path to Lean Maintenance.

2. Pilot Kaizen Events critiqued with each Action Team
3. Reviews of Pilot Events and Lessons Learned presented to the rest of the maintenance and maintenance support organizations

Phase 4 Lean Mobilization (6 Months to 1 Year)

Lean Tools are unleashed within the Maintenance Operation during Phase 4. It is in this mobilization phase that the leadership qualities of the Lean PM and the Lean Project Team are most needed. In order to sustain the progress made in Phase 3, constant push and positive reinforcement by the PM and Project Team members will be required. Any evidence of slippage must be corrected immediately before it spreads throughout the maintenance and support organizations. Course correction needs to be accomplished in a positive fashion. Placing blame must absolutely be avoided.

Getting the rest of the maintenance and maintenance support infrastructure converted into empowered, self-directed work teams that are implementing Lean can be an overwhelming challenge. It is a very real possibility that even the members of the project team will experience frustration, discouragement and an urge to abandon the effort. The Lean PM should be in constant communication with all team members as a group and with each team member as individuals. Periodic motivational sessions that are addressed by a respected member of upper management are strongly encouraged. The qualities and attributes (refer to Section 4.1.1) of the Lean PM are never more in need than during Phase 4.

Phase 4 Activities:

1. Establish office space or area as "Lean Maintenance Central" for ongoing status display and central transformation control.
2. Maintenance staff and maintenance support activity's staff converted to individual Action Teams.
3. 5-S deployed early as method for Action Team involvement.
4. Visual cues used extensively.

5. Action Team leads meet and discuss issues as a group.
6. Overall managerial/supervisory roles change as action teams develop identities.
7. Organization changed to one of customer-focused emphasis (production department is immediate customer; product consumer is the ultimate customer).
8. Pay-for-performance system deployed as action teams mature and measurable improvements realized.
9. Inter- and intradepartment communications opened, exercised and reinforced.
10. As maintenance-staffing requirements are reduced through improved efficiencies, reassign selected staff to new, predefined positions. Allow normal attrition to reduce staff size. Do not terminate anyone for other than flagrantly poor performance.
11. Continuous learning and improvement is everywhere.

Phase 5: Lean Expansion (4 Months to 1 Year)

This phase takes Lean Maintenance outside of the plant walls to the supply chain and specialized contractors. During this phase the maintenance department creates connected value streams for maintenance stores and applicable maintenance tasks. Depending on the maturity of the Total Productive Maintenance (TPM) system, this phase can take as little as one to two months to complete in the very best case scenario.

The objectives are to minimize maintenance inventory and cost of inventory while continuing to meet equipment reliability needs and to perform TPM processes in the most efficient (time and cost) manner possible.

Supply Chain—In the past, one of the objectives of the maintenance storeroom was to have on hand as many spare parts as possible, just in case they might be required. But the increasing cost of inventory is making that practice obsolete. Over half of existing inventory can be eliminated by using your CMMS to schedule, based on maintenance action scheduling, when repair parts and consumables will be needed. Then your supplier is provided with specific delivery requirements to meet those maintenance schedules (JIT delivery). Nearly all PM related parts and consumables could be removed from your storeroom. Incoming items can go directly to a staging area (much smaller than the storeroom space previously occupied) for issue to the following week's PM work orders.

Accurate equipment inventories in your CMMS aid in accurately identifying repair parts requirements. By providing these requirements to your supplier, you can contract for minimum lead-time delivery (JIT) of those parts. Now, instead of carrying sufficient parts for six month's usage, minimum lead-times (for example—one week) will allow you to reduce that level to one week's usage. You've just achieved a 26:1 reduction in on-hand parts inventory.

The trap that you must avoid here is that you also must reduce inventory cost. If all you have accomplished is moving 96% of your inventory from your storeroom to your supplier's storeroom, the cost simply shifts to your supplier. And you can bet that he is going to pass that cost on to you—hence no reduction in inventory cost. Purchasing personnel must work effectively with suppliers, doing much more than merely providing them with lead-time requirements. Providing inventory demand estimates and usage data to your supplier helps him to define his own inventory levels, which he can then provide to the manufacturing plant that produces that particular maintenance inventory item. And, if that manufacturing plant happens to be a Lean Manufacturer, you have just helped to create a completely Lean parts supply loop. Purchasing must encourage suppliers to practice Lean thinking or identify suppliers already employing Lean practices.

Standardizing suppliers as well as consumables and material items issued from the storeroom is also required during this phase. Especially in plants with decentralized storerooms, it is more than likely that the identical repair part is procured from multiple suppliers. Consolidating the purchasing effort can be used to identify the most cost-effective supplier and standardize with that single supplier.

In order to standardize material items and consumables issued by the storeroom you will need to select a starting point. Safety supplies are a good choice. For one thing, the number of employees who use safety supplies provides a good pool of end users from which to form a cross-functional team. Here we will use safety glasses as an example. When bringing the team members together, ask each end user to select one pair of safety glasses and provide a list of features that he or she liked about the glasses as well as those he or she did not. Compiling these listed features creates a single list from which a choice of one or perhaps two safety glasses that meets all, or nearly all the desired features and avoids all, or nearly all the undesirable features can be made. Many manufacturing plants carry ten or more styles of safety glasses. Reducing that number to one or two lowers the inventory level, which in turn reduces inventory costs. Not all items require this kind of selection process; in most cases the storeroom supervisor or purchasing employee can quickly identify desirable and undesirable features and narrow the choices.

TPM Processes—In the Total Productive Maintenance environment, much use is made of Predictive Maintenance (PdM) techniques and equipment and Condition Monitoring/Condition Testing. Many of the techniques used require highly skilled operators using very expensive test and monitoring equipment. If your plant contains enough production equipment to produce a positive trade-off comparison for the cost of test equipment (including repair and calibration costs), the cost of training operators and the cost of training Maintenance Engineering staff in analysis, over three years when compared to the cost of contracting for the testing and analysis over three years, then it is efficient to perform those processes in-house. Because of the technology involved with most PdM and CM/CT techniques, much of the equipment utilized can become obsolete in 40 to 48 months, which is why the cost trade-off is performed over a 36-month period. If the trade-off comparison is negative, the services should be contracted.

Phase 5 Activities:
1. Obtain inventory item lead-time requirements and demand/usage data from CMMS and work with suppliers for JIT delivery.
2. Standardize material and consumable items and suppliers
3. Standardize suppliers of identical repair parts
4. Perform trade-off analysis of the costs of performing PdM and CM/CT processes; contract out for those with negative results.

Phase 6: Lean Sustainment (for the life of the company)

The Maintenance Department and Maintenance Support Sectors gain full ownership of Lean Maintenance in this phase. Lean thinking is being sustained and every employee within these areas is actively involved in continuously improving their processes. New employees receive an initial indoctrination in Lean practices and the Action Team to which they are assigned provides the remainder of their Lean training.

The Lean PM and Project Team have completed their "tour-of-duty." A Lean Leader for Maintenance and Maintenance Support functions should be appointed. A single point of leadership is an important aspect of sustainment. The Lean PM can stay on as the Lean Leader, but it should be expected that his or her knowledge and experience will be aggressively recruited for the plant-wide Lean transformation.

Daily leadership of the Lean Maintenance organization passes to the organizational leaders. Their performance evaluations will look closely at their Lean leadership efforts. Lean leadership characteristics take on more of a counseling role than a problem-solving role. The organizational leadership assists the progress and sustainment of Lean Maintenance by removing obstacles that Lean action teams face. They ensure that action team leaders are knowledgeable and active in their roles. They ensure that the action teams remain empowered.

4.3 LEAN MAINTENANCE TRANSFORMATION KICK-OFF MEETING

Once the Project Team indoctrination has been completed, the PM and Project Team should prepare for a Lean Maintenance Transformation kick-off meeting. It is at this point that the Project Team members begin to assume their leadership roles. To prepare for the kick-off meeting this group should hold a brainstorming session, facilitated by the Lean PM, to determine/develop:

- Kick-off Meeting Agenda (Minimum Information):
- Description of Lean and How Lean Transformation Will Take Place
- Assignment of Action Team Members
- Definition of Each Action Team's Responsibilities
- Road Map of the Journey
- List of Attendees (normally managers, supervisors and recognized leaders within the various work centers—and Action Team Members)
- Define What Things the Meeting Should Accomplish (Use a Checklist)

It is important to remember while developing the agenda and the checklist of "Things the Meeting Should Accomplish" that the purpose of the kick-off meeting is to educate and to sell. Selling the managers involves defining expected gains in efficiency, improved reliability, cost reductions and similar accomplishments. At the supervisor and maintenance mechanic/technician level, interest is higher in things such as job security, improvements in pay and benefits, improvements in working conditions and related accomplishments expected from adopting Lean Practices. An example of a comprehensive Kick-Off Meeting Checklist is shown in Table 4-1.

The checklist that you develop for your kick-off meeting can contain more detail than that in Table 4-1, but it should not contain any less. This meeting introduces Lean to the leaders of the maintenance operation and unintentionally leaving out key information can lead to doubt, unenthusiastic support and even overt attempts to make the transformation fail. You cannot provide too much information!

Table 4-1
Kickoff Meeting Checklist

Upper Management Representative's Intro emphasizing commitment	✓
Define Lean Maintenance	
It is—Elimination of Waste—"EVERYWHERE"	✓
Value Added Activities	✓
Customer Focused	✓
Problem Solving	✓
Empowering of Shop Floor Operators and Maintenance Techs	✓
It isn't Fault-Finding of People—People aren't problem, they're problem solvers	✓
Down-Sizing—No lay-offs; pay-for-performance	✓
Why Lean Maintenance?	
Maintenance operation largest supporter of production; must be prepared before plant and production can go Lean	✓
Gain Competitive Advantage	✓
Improve Quality	✓
Lower CPU (Cost per Unit)	✓

Lean Accomplishments: Wrench Time (time actually performing Maint.) ↑50%
   Equipment Reliability ↑25%
   Emergency Work ↓75%
   Maintenance contribution to CPU ↓30%
	✓
	✓
	✓
	✓

Lean Transformation	
Introduce Project Team—Explain Make-up	✓
Road Map—Phase 1; Explain and describe results	✓
Phase 2 through 6—Describe each. Show POA&M, duration	✓
Emphasize Phase 6 ◊ forever (continuous improvement)	✓
Describe what Action Teams are and explain their activities and responsibilities	✓
Why there are only 3 or 4 Action Teams to start	✓
Assignments to initial Action Teams	✓
Phase 4 ◊ All hands formed into Action Teams	✓
Discuss next phase (2)—Education; the schedule, attendees	✓
Invite Questions—now or in private, one-on-one sessions with Lean PM or Project Team	✓

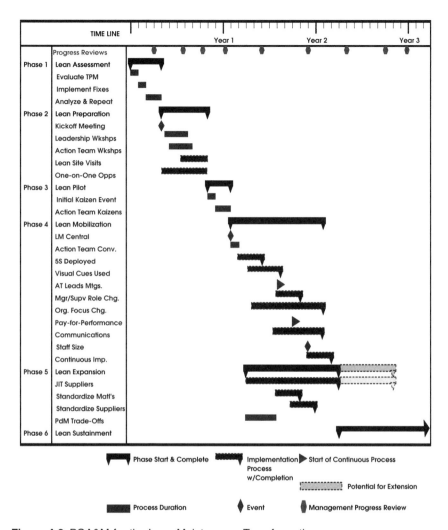

Figure 4-2 POA&M for the Lean Maintenance Transformation

4.4 PHASE 1: DEVELOPING THE POA&M AND THE MASTER PLAN

Following the process of creating the kick-off meeting agenda and preparing the presentation, which includes an outline of the Lean Maintenance Journey, a more detailed Plan of Action and Milestones (POA&M) that schedules the processes and milestones of the journey's roadmap should be created. The POA&M should indicate generally expected results and not specific events such as Kaizen Blitz events. It should indicate the expected duration and completion dates of each phase of the Lean transformation and show the major accomplishments, such as "5-S Fully Deployed" and "JIT Suppliers Identified and Contracts Signed" within each phase. It should also include a schedule of "formal" progress reviews (as opposed to weekly status reports).

Phase duration and event or process completion and duration schedules should be based on the Lean PM's and Project Team's knowledge and understanding of the maintenance organization's present level of effectiveness, as well as capability and motivation potential. Aim for accurate projections, but err on the side of conservatism. The POA&M will be the gauge for progress, and more enthusiasm is generated when progress exceeds planned than the opposite. An example of a POA&M for the Lean Maintenance Transformation is illustrated on the previous page.

With the development of the schedule or POA&M you now have all of the elements of the Master Plan. They include:

- Lean Maintenance Mission Statement and Vision
- Lean Maintenance Objectives, Goals and Targets
- Phase 1 Results (TPM Evaluation)
- Kick-Off Meeting Agenda and Checklist
- POA&M

All of the elements should be organized into a single document "The Master Plan for a Lean Maintenance Transformation." The Master Plan should now be submitted to upper management for review and approval. At the time of submission you should also attempt to schedule a meeting with management to review the master plan. If at all possible, the review should take place within a week of submission. Your objective is to obtain management approval of the plan so that the kick-off meeting can be scheduled and Phase 2 (Education) commenced as soon as possible. The Lean Maintenance Transformation is a long journey and delays will only make it longer.

Ensure that you are fully prepared for the master plan management review. The review meeting is yet another opportunity to sell and obtain commitment. Don't let it go by without making the most of it. Stress the gains to be made with a successful transformation. If management originated the "Lean" initiative, be sure to pay attention during the review. Management would not undertake such a transformation without knowing all of the details. Demonstrate that you have the knowledge as well as the leadership qualities to make the journey successful.

5

Launching the Master Plan (POA&M)

5.1 THE SEQUENCE OF EVENTS

Following the completion of the Kick-Off Meeting, there should be approximately one week set aside before the series of workshops begins. During this week encourage feedback from those attending the meeting. Schedule "office hours" for one-on-one sessions or group meetings if attendees desire. It is important to set minds at ease and eliminate any fears that may exist so that when the workshops do get underway the attendees can devote their complete attention to them without any preoccupation.

Although it appears that we are just beginning our journey to becoming a Lean Maintenance Operation, we have actually come a long way. Let's briefly review what we have accomplished thus far:

Pre-Lean
- Determine Need for CMMS
- Define CMMS Requirements/Evaluate Vendor Software
- Select CMMS Vendor
- Install CMMS/Train IT and Maintenance Users
- Implement CMMS (populate databases)
- Implement Proactive Maintenance Culture
- Upgrade Equipment Reliability
- Implement Total Productive Maintenance
- Establish TPM Performance Metrics

} Performed Simultaneously

Lean Preparations
- Select Lean Project Manager (PM)

- Lean PM Learning/Training
- Assignment of Lean Project Team
- Lean Planning

Lean Transformation
- Lean Assessment (Phase 1)
- TPM Effectiveness Evaluation and Upgrade (as required)
- Develop Lean POA&M and Master Plan
- Submit Master Plan for Management Approval
- Lean Kick-Off Meeting/Follow-up
- Lean Preparation (Phase 2) Education ← We are just starting this phase.

Phase 2, *Lean Preparation* and Phase 3, *Lean Pilot* together constitute going public with the Lean Maintenance transformation or the Launching of Lean.

5.1.1 Phase 2—The Lean Preparation Phase (Education)

The Preparation Phase is predominantly an education period: Two series of workshops, one a series for management and the other for the shop floor level supervisors, maintenance technicians and maintenance support personnel (maintenance administration, maintenance engineering, maintenance storeroom, production line supervisors and operators) and any others designated by the Project Team. In order to keep attendance at the workshops at a manageable level, they may need to be repeated two or more times. Maximum workshop class size should be limited to no more than 25. The initially designated Action Teams will need to attend the workshops in team units because of the practical exercises performed in the workshops.

The content of the workshops should be patterned after the Project Team training sessions (refer to Section 4.1.3.1).

The Action Team Lean Principles and Techniques Workshops must be expanded to include actual exercises in the application of the following Lean Implementation Tools.

5.1.1.1 5-S (Visual)

Application of the 5-S Tool focuses on effective workplace organization and standardized work procedures. 5-S simplifies your work environment and reduces waste and non-value activity while improving quality efficiency and safety.

1. Sort (Seiri)—The first S focuses on eliminating unnecessary items from the workplace. An effective visual method to identify these unneeded items is called red tagging. A red tag is placed on all items not required to complete your job. These items are then moved to a central holding area. This process is for evaluation of the red tag items. Occasionally used items are moved to a more organized storage location outside of the work area while unneeded items are discarded. Sorting is an excellent way to free up valuable floor space and eliminate such things as broken tools, obsolete jigs and fixtures, scrap and excess raw material. The Sort process also helps prevent the JIC job mentality (Just In Case).

2. Set In Order (Seiton)—The second S focuses on efficient and effective storage methods.

   You must ask yourself these questions:
   - What do I need to do my job?
   - Where should I locate this item?
   - How many of this item do I need?

   Strategies for effective *Set In Order* are painting floors to outline work areas and locations, create shadow boards and install modular shelving and cabinets for needed items such as trash cans, tools and toolboxes, clipboards and anything needed on a recurring basis. How much time is wasted every day looking for a frequently used item? Each item should have a specific location where all employees can find it. "A place for everything and everything in its place."

3. Shine (Seiso)—Once you have eliminated the clutter and junk that has been clogging your work areas and identified and located the necessary items the next step is to thoroughly clean the work area. Daily follow-up cleaning is necessary in order to sustain this improvement. Workers take pride in a clean and clutter-free work area and the *Shine* step will help create ownership in the equipment and facility. Clean shop and production areas make faults stand out more clearly. Workers will begin to more quickly spot changes in equipment and services such as air, oil and coolant leaks, repeat contamination, vibration, broken items and misalignment. These changes, if left uncorrected, could lead to equipment failure and loss of production. Both add up to impact your company's bottom line.

4. Standardize (Seiketsu)—Once the first three of the 5-Ss have been implemented, you should concentrate on standardizing best prac-

tices in each work area. Shop-floor maintenance technicians are the best source for the development of such standards. They are a valuable but often overlooked source of information regarding their work and work spaces.

5. Sustain (Shitsuke)—This is by far the most difficult *S* to implement and achieve. Human nature is to resist change and more than a few organizations have found themselves with a dirty cluttered shop a few months following their attempt to implement 5-S. The tendency is to return to the status quo and the comfort zone of the "old way" of doing things. Sustain focuses on defining a new status quo and standard of workplace organization.

5.1.1.2 Standardized Work Flow

This refers to work process standardization. There are normally two, or occasionally three, starting points for a maintenance work order—a CMMS generated work order (planned), a Production generated work request and occasionally Maintenance Engineering (special evaluation, etc.). Where the work request goes next starts the beginning of the Work Flow process. If some go to the maintenance shop and others to the Planner and Scheduler, the process is not very standardized. And if emergency work requests are not always written (i.e., phoned to maintenance) a valuable source of information for analyzing equipment failure causes is lost.

Typically various communications will follow, sometimes requiring input from several parties. There are many ways to perform a given repair, and each alternative calls for different methods, parts and costs. Also, opinions vary widely about priority, so at this stage, a maintenance supervisor or worker familiar with the equipment often makes a field check to evaluate the work needed and identify the best repair method.

Once a method is approved, the planner/scheduler determines work steps, safety needs, permits, tools, equipment and material. After all job elements are informally defined and materials procured, the planner determines the crew size, trades or skills and estimates the job time. The work order then is entered into the ready-to-work backlog for assignment, by priority, to a maintenance worker.

Scheduling, assigning and completing work orders—The next step is to schedule the job based on priority, assigning as much as a week's work to one individual according to the skill required and crew member's experience. After assignment, the individual executes the work to specification and to the time scheduled. The supervisor's follow-up to determine completeness and quality generally is limited, and the requester seldom is notified that the work is done unless the job was urgent. The worker reports the time and work performed. This information too often is inaccurate or incomplete, and the clerk posts this suspect labor and material information on the equipment record. Finally, the work order is closed and removed from the backlog.

Measuring results—Few departments actually measure results. Various control reports and trend charts that managers can use to track performance and identify needed corrective action generally are not available, or they are used infrequently because managers know the information is unreliable. Key report control indicators including performance, coverage, delays and cost per standard hour produced might be unavailable.

All of these variances and player's choices must be mapped, evaluated, optimized, standardized and deployed. A very abbreviated standardized workflow is illustrated in Figure 5-1.

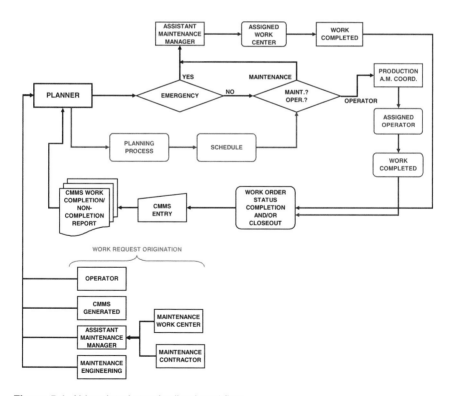

Figure 5-1 Abbreviated standardized workflow

5.1.1.3 Value Stream Mapping

This is a powerful tool for "seeing" a process, identifying the non-value adding components and recreating the process as a value stream. The mapping process employs several standard map symbols that were created for manufacturing processes. They are usable for the maintenance operation as well, but the important thing in visual stream mapping is for the map to be easily understood, so if you are more comfortable using symbols that you devise, use them. Some of the more common symbols are shown here in Figure 5-2.

Another often-used process-mapping technique dating back to the early 1900s is the original system developed by Frank Gilbreth, which is still very useful. The Gilbreth approach is highly visual and discriminates between waste and value-added activity very clearly. It is also simple and easily used by even untrained groups. Annotate each symbol by describ-

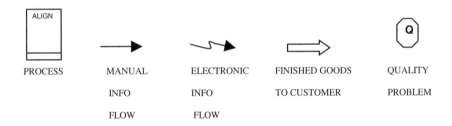

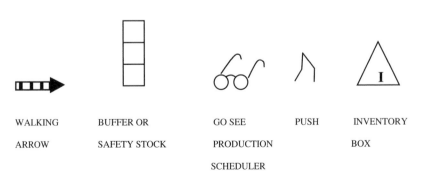

Figure 5-2 Common Symbols

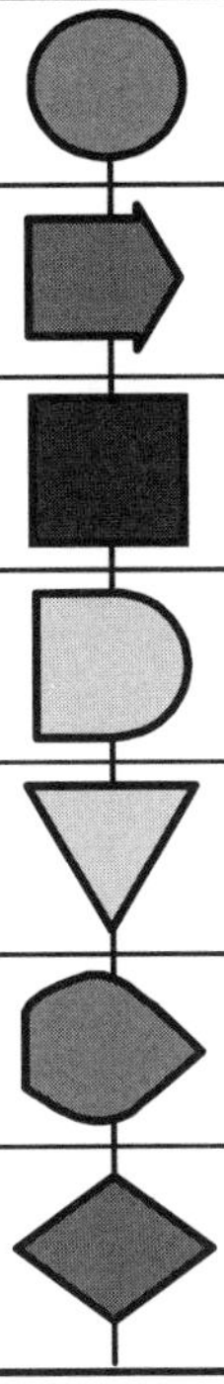

SYMBOL	MEANING	ACTION	EXAMPLES
	OPERATION	ADDS VALUE	CUT, SOLDER, MEASURE, ETC.
	TRANSPORT	MOVE SOME DISTANCE	CONVEY - BY HAND-CARRY, FORKLIFT, ETC.
	INSPECT	CHECK FOR DEFECTS	DIMENSIONAL VISUAL, ETC. INSPECTION
	DELAY	TEMPORARY DELAY OR HOLD	WAIT FOR EQUIP. SHUTDOWN, ETC.
	STORAGE	FORMAL WAREHOUSING	WAREHOUSE, STOREROOM, OTHER STORAGE
	HANDLE	TRANSFER OR SORT	RE-PACKAGE, RETURN TO STORES, ETC.
	DECIDE	MAKE A DECISION	APPROVE/DISAPPROVE, TAKE OFF-LINE, ADJUST OR REPLACE, ETC.

Figure 5-3 Gilbreth Approach

ing the event as concisely as possible and indicating the time required, as in Figure 5-3. Appendix B contains an example of this mapping technique.

Whatever process-mapping system you decide to use, its application is the same. It employs the following 8-Step Value Stream Mapping and Future State Creation plan:

1. Select the process to be mapped and study/analyze it carefully.
2. Map the process' existing steps.
3. Reanalyze by examining each map symbol and attempting to "drill down" to additional process steps within each mapped step. Continue until the team agrees that all steps of the process have been mapped. This results in the present-state map.
4. Analyze the present state map to identify all non-value adding activities.
5. Remove the non-value adding activities or develop value-adding "work-arounds" and remap the process. Create a listing of all of the

actions needed to remove the non-value adding activities as well as any value added work-arounds developed.

6. Reanalyze the new map for workability and additional non-value adding activities. Continue until the entire team agrees that the process is now workable and consists of only value added activities and "impossible to remove" non-value adding activities. The result is the process' future state map. The listing of the actions needed to remove the non-value adding activities as well as any value added work-arounds developed constitutes the steps of an action plan for modifying the selected process.

7. Write up your action plans and submit them to management for approval.

8. Implement the process' action plan in accordance with approval guidelines.

5.1.1.4 Just-in-Time (JIT) and Kanban "Pull" System

Although this tool has many more applications on the production line than in the maintenance operation, there are still some very useful features that can "slimdown" a major maintenance process—scheduling or throughput. JIT and Pull are really facilitators of continuous flow in production terms. How does that translate to maintenance operations? The goal in production operations is continuous flow such that production can (potentially) run at full capacity. What is full capacity in maintenance operations? It is performing the right amount of maintenance required to meet the production schedule that is approved by the customer who, in most cases, is production. *Approved* means approval before the maintenance job starts (production wants and needs this work to be done) as well as approval after job completion (production is on-line and happy with the quality of the work).

How much is this maximum amount of maintenance? In terms of work management the design capacity of the maintenance process is equal to the total number of man-hours available for executing maintenance work. This design capacity is used for scheduling maintenance. Scheduling 100% of available labor hours will enable the "maximum amount of maintenance," or full design capacity, to be achieved. Schedule anything less and full design capacity can never be reached. Managers and planners can provide input for keeping some level of unused capacity in reserve for high-priority work (not for emergency work) and unplanned absences. Maintenance and production are thereby empowered to fill in the remaining capacity on the day before execution. The result is (should be) a realistic schedule allowing for long-term planning of high-priority work with enough flexibility to accommodate some level of operational change.

With a schedule for 100% utilization of capacity, 100% (schedule) compliance means maximum maintenance throughput. This level of compliance should be the goal of everyone. Thus, schedule compliance is the key performance indicator (KPI) for the work management process. Anything less than 100% schedule compliance should readily highlight problem areas that can be targeted for resolution.

5.1.1.5 Jidoka (Quality at the Source)—Poka Yoke (Mistake Proofing)

As this relates to manufacturing, *jidoka* means quality is manufactured in by the process and not inspected in. This kind of quality control can certainly be applied to maintenance processes. Trained, skilled and qualified maintenance technicians should be performing, or directly supervising, every maintenance procedure. He or she should be using the proper tools, have the correct repair parts and consumables (lubricants, cleaning agents, coolant, etc.) and be working in the proper environment. Pride in workmanship should be a very highly rated factor in job performance.

5.1.1.6 Shewhart Cycle (PDSA)

The Shewhart Cycle of *Plan—Do—Study—Act* is the control process for executing Kaizen Events. Dr. Walter Shewhart, one of W. Edwards Deming's mentors, is primarily known for development of statistical process control methods, but he also developed the Shewhart Learning and Improvement Cycle, combining both creative management thinking with statistical analysis. This cycle contains four continuous steps: Plan, Do, Study (or Check) and Act. These steps (commonly referred to as the PDSA or PDCA cycle), Shewhart believed, ultimately lead to total quality improvement. The Cycle is based on the premise that continual evaluation of management practices, as well as the willingness of management to adopt new, and disregard unsupported, ideas are keys to the evolution of effective management and a successful enterprise.

The PDCA cycle stresses experimentation and observation as the means of discovering truth:

- In the *Planning* stage, the problem is recognized and analyzed, and possible solutions formulated.

- In the *Doing* stage, the most likely or effective solution is implemented in a test site.
- The *Study* or *Check* step is used to compare results of the test solution and the original method to see if there are real improvements.
- *Acting* involves replacing the old method with the successful solution.

You can then return to the beginning of the cycle to explore other possible problems and continually strive for new levels of improvement.

The search for common causes is just one of the many arenas in which the PDCA cycle can be used. Most generally, it is used to guide overall process improvement, of which searching for common causes might be just one element.

The PDCA Cycle calls for both creative thinking and analytical thinking, each essential to process improvement. Creative or divergent thinking encourages many ideas to be considered and new possibilities to be uncovered. Creativity is an important factor because it can break through flawed paradigms and see beyond the current way of thinking about a process. But creativity must be tempered by critical analysis or convergent thinking that brings the scattered pieces back together in a workable form.

One often used method during the *P* (Planning and Analysis) stage is the creation of Cause and Effect diagrams. The Cause and Effect diagram (often called the fishbone diagram because of its appearance) is the brainchild of Kaoru Ishikawa, who pioneered quality management processes in the Kawasaki shipyards, and in the process became one of the founding fathers of modern management. The Cause and Effect diagram is used to explore all the potential or real causes (or inputs) that result in a single effect (or output). Causes are arranged according to their level of importance or detail, resulting in a depiction of relationships and hierarchy of events. This can help you search for root causes, identify areas where there may be problems, and compare the relative importance of different causes.

Causes in a Cause and Effect diagram (as in Figure 5-4) are frequently arranged into four major categories. While these categories can be anything, you will often see:

- people, processes, materials and equipment (recommended for manufacturing and maintenance)
- equipment, policies, procedures and people (recommended for maintenance)

These guidelines can be helpful but should not be used if they limit the diagram or are inappropriate. The categories you use should suit your needs. The following are the generalized steps for creating a Cause and Effect diagram.

STRUCTURE OF CAUSE AND EFFECT (FISHBONE) DIAGRAMS

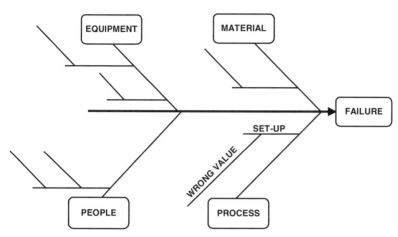

Figure 5-4 Cause and Effect Diagram

1. Be sure everyone agrees on the effect or problem statement before beginning.
2. Be succinct.
3. For each effect, think what could be its causes. Add them to the tree.
4. Pursue each line of causality back to its root cause.
5. Consider grafting relatively empty branches onto others.
6. Consider splitting up overcrowded branches.
7. Consider which root causes are most likely to merit further investigation.

5.1.2 Lean Pilot (Phase 3)

5.1.2.1 Selecting the Project

It is extremely important that the initial Kaizen Event performed by the Primary Action Team has dramatic results. The full impact of the effectiveness of adopting Lean practices is best made when large gains are realized. One method for choosing your first Kaizen Event process employs the Pareto Principle.

Vilfredo Pareto (1848–1923) was an Italian economist who, in 1906, observed that 20% of the Italian people owned 80% of their country's accumulated wealth. Over time his observation has been applied to a variety of situations. It has come to be referred to by many different

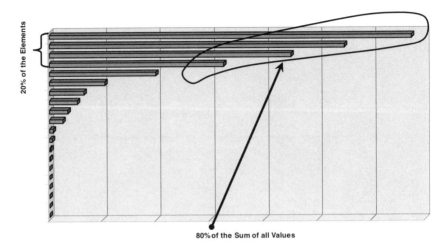

Figure 5-5 Pareto Principle

names, including *Pareto's Principle,* the *80-20 Rule* and the *Vital Few and Trivial Many Rule.* Called by whatever name, this mix of 80%-20% illustrates that the relationship between input and output is not balanced. In a management context, this rule of thumb is a useful tool that applies when there is a question of effectiveness versus diminishing returns on effort, expense or time. It can be applied for selection of processes that account for the more significant uses of labor, time, production downtime, cost or any other parameter of your choosing. Graphically, the Pareto Principle looks like Figure 5-5.

To create a Pareto Chart:

1. Select the items (problems, issues, actions, publics, etc.) to be compared.
2. Select a standard for measurement.
3. Gather necessary data.
4. Arrange the items on the vertical axis in a descending order according to the measurements you selected.
5. Draw a bar graph where the length is the measurement you selected.

Practical Applications of the Pareto Principle—Some examples about the allocation of time, effort, and resources are the following:

- Costs: To reduce costs, identify which 20% of equipment items are using 80% of the resources. If members of this segment are not top profit generators, consider charging them for the resources they consume or shift services away from this sector.

- Personal Productivity: To maximize personal productivity, realize that 80% of one's time is spent on the many trivial activities. Analyze and identify which activities produce the most value to your company and then shift your focus so that you concentrate on the vital few (20%). What do you do with those that are left over? Either delegate them or discontinue doing them.
- Profits: To increase profits, focus attention on the vital few (top 20%) by first identifying and ranking customers in order of profits and then focusing sales activities on them. The 80-20 Rule predicts that 20% of the customers generate 80% of the revenues, and 20% yield 80% of the profits, but these two groups are not necessarily the same 20%.

More Examples of the 80-20 Rule:

- 80% of a manager's interruptions come from the same 20% of the people
- 80% of a problem can be solved by identifying the correct 20% of the issues
- 80% of an equipment budget comes from 20% of the items
- 80% of benefit comes from the first 20% of effort
- 80% of what we produce is generated during 20% of our working hours
- 80% of your innovation comes from 20% of your employees
- 80% of your staff headaches come from 20% of your employees

Selection and prioritizing projects for Kaizen Event candidacy, whether you apply the Pareto Principle or some other method of prioritization, requires information from your CMMS. Some CMMS-generated reports that are useful in pinpointing waste and processing problem areas include:

Equipment

1. Equipment Maintainability—Total downtime ÷ number of downtime occurrences (by equipment)
2. Overall Equipment Effectiveness—Availability × Utilization × Quality

Maintenance Processes

1. Overall Measures (Department)—percent of effort on Breakdown, Corrective, Predictive and Preventive Maintenance
2. Overall Measures (by Craft)—percentages as previous
3. Percentage work orders delayed due to Planning and Scheduling—delays to parts availability, equipment availability, other delays due to poor weekly and daily scheduling processes

4. Interruptions to the schedule—measure by a number and category (e.g., production, maintenance, etc.)
5. Work Order Life by Priority—number of work orders by age within each priority level
6. Callbacks—by equipment and by work center
7. Estimating Efficiency—actual hours worked divided by the estimate
8. Percentage of work orders generated from preventive maintenance inspections/services
9. PM Compliance—percentage of scheduled PMs performed within 10% of the frequency
10. Overtime Percentage of Total Labor
11. Average Work Order Life
12. Filtered Backlog—backlog sorted by:
    • Area of Maintenance (operational, shutdown, technical change, etc.)
    • Equipment
    • Originator
    • Work Order Type (Safety, Capital Improvements, Preventive, Predictive, Corrective, Shop Repair, Breakdown)
    • Non-compliant Work Orders (no priority, erroneous work order codes, insufficient data, scheduled/unplanned work orders, closed corrective work orders without failure codes, work orders with little or no completion comments)

Using the Pareto Principle and CMMS reports or any other viable method, a prioritized listing of Kaizen Event candidate processes should be generated. The priorities established should then be followed in the selection of Phase 3 Action Team projects.

5.1.2.2 The Pilot Kaizen Events

The maintenance operation work process for the primary Action Team's Kaizen Event is selected from the top of the candidate's list that you have just created. But, before beginning it's prudent to perform one more analysis of the selected process by asking ourselves (as a team) a few questions:

• Do we have sufficient knowledge of the process?
• Do we need any specialized talent/knowledge not presently on the team? (or should it be handed-off to another team?)
• Do we believe the process can be significantly improved?

Planning (Plan, Analyze & Develop Solutions)	✔
Analyze and develop present state map	
Drill down for additional steps	
Team agreement on present state map	
Analyze for and apply appropriate solution methods	**(Y or N)**
Flow (Develop Future State Map)	
Cause and Effect	
Visual Controls	
Standardized Work	
Mistake Proofing	
Training Needs Analysis	
Doing (Implement Appropriate Test Solutions)	✔
Apply solutions from Planning	
Apply 5-Ss as appropriate	
Study/Check (Compare results to original state)	✔
Select effective solutions to keep	
Deactivate ineffective solutions	
Remap future state	
Begin listing selected, effective solutions	
Acting (Implement revised future state across entire process)	✔
Apply effective solutions selected above	
Go back - to Planning and repeat PDSA (refinement)	

Figure 5-6 PDSA Checklist for a Kaizen Event

- Do we believe that process improvement will disrupt other plant processes? Substantially? (This may result in selecting an alternate)
- Can we perform the PDSA (PDCA) cycles in one week or less?

When the Action Team has satisfied itself, as well as the project team, that it is practical to proceed, the event begins. Here again, there is a correct sequence to follow (see the checklist in Figure 5-6) in order to avoid stall-outs and wrong turns.

    If possible, it will facilitate follow-on training to videotape as much of this process as feasible. But following this process' completion, the next step is to critique the Kaizen Event with the primary Action Team. The first portion of the event's review should solicit the team's opinions and expressly find out what things they would do differently. The debriefing should lavish the praises, but also pinpoint errors or flawed/weak elements of the event and then repeat the lavish praises. Following the critique, the event synopsis should be taken to the rest of the maintenance and maintenance support organization. The lessons learned—good and bad—should be learned by everyone.

Following the critique and the presentation of the event "Lessons Learned" to the rest of the organization, the Follow-on Action Teams (simultaneously) will follow the same process in selecting and executing their own Kaizen Events. These will be monitored, critiqued and additional "Lessons Learned" should be passed on to the rest of the maintenance and maintenance support organization.

6

Mobilizing and Expanding the Lean Transformation

6.1 MOBILIZING LEAN IN THE MAINTENANCE ORGANIZATION (PHASE 4)

Lean Mobilization and Lean Expansion move the Lean Maintenance transformation into the hands of the Maintenance Department (Phase 4) and the Maintenance Support Organizations (Phase 5). The training provided during the Lean Preparation and Lean Pilot Phases (2 and 3) has equipped everyone with adequate knowledge of the principles and application of Lean so that they can indeed become self-directed teams.

What Occurs in Phase 4:
All Maintenance Department Formed into Action Teams
✓ 5-S/Visual Cues Campaign
✓ Autonomous Operator Maintenance Instituted
✓ Action Team Leaders Share Knowledge
✓ Complete Maintenance Department Lean Mobilization

CHANGES BROUGHT ABOUT BY MOBILIZATION
Roles of Management and Supervision
Maintenance Department Organizational Focus

6.1.1 Teams and Activities in Phase 4

Following the Lean Pilot Events and the resultant training, the maintenance department and support operations should be formed into teams with the objective of implementing Lean practices and carrying out Kaizen Events within their areas of cognizance and responsibility. When the Primary and Follow-on Action Teams were formed for the pilot events, membership was predicated primarily on personal attributes such as knowledge and skill levels, leadership qualities, self-starting characteristics and enthusiasm for improvement. Team formation during the mobilization phase will be based on work area, equipment responsibility, craft and departmental assignments. Following the maintenance-wide formation of action teams, the Primary and Follow-on Action Teams will provide support to assist in their mobilization and are then disbanded and their members assigned to teams based on these new criteria. The intent here is to have the teams performing their day-to-day activities, as much as is feasible, together as a team, as well as their Lean Improvement activities. Ideally, organizational changes enacted during Pre-Lean and Lean Preparations will facilitate this kind of team formation. Otherwise there may need to be some additional changes in the organization's structure to enable effective team assignments. Action Team characteristics include:

- Team activities are task oriented and designed with a strong performance focus.
- The team is organized to perform whole and integrated tasks hence requiring multi-department membership.
- The team should have defined autonomy (that is, control over many of its own administrative functions such as self-planning, self-evaluation and self-regulation all with limits defined); furthermore, members should participate in the selection of new team members.
- Multiple skills are valued; this encourages people to adapt to planned changes or occurrence of unanticipated events.

Converting the maintenance department shop floor into self-directed action teams is a major undertaking with many opportunities for disaster. The conversion can't be completed overnight. It needs to be planned for and carried out as part of a well-thought-out process. Generally there are four steps to perform the conversion. (see Figure 6-1)
Preparation

- Establish work areas and work objectives
- Define membership by craft, department and skill levels
- Define team level of authority/autonomy

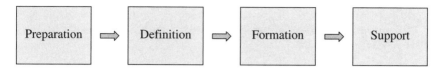

Figure 6-1 Four Steps to Perform Conversion

Definition

- Define team task
- Select team members
- Establish support infrastructure and material requirements

Formation

- Make team assignments and publish
- With the team, define its boundaries
- With the team, refine task definition
- With the team, define leadership and member roles

Support

- With the team, negotiate aspects of performance goals
- Provide process assistance as needed to promote team synergy
- Assist in development of opportunities for improvement activities
- Establish processes for ongoing support

Some essential knowledge for action team members—and leaders—to possess includes:

- It's difficult to achieve a coordinated team effort when people leave their stations—stray into someone else's area–or just get sloppy and let things slip through the cracks.
- You must be clear on what's expected of you—duties, standards of performance, time frames and deadlines you're supposed to meet. Specifically, what moves are you supposed to make? In what sequence? What territory are you supposed to cover? How should you interface with others on the team?
- Teamwork, by definition, implies interdependence. What you do affects others. Some people in the unit depend on you for their success, their effectiveness. What you fail to do can cause them to fail. Chances are if you fall down on the job, you'll pull others down with you. If you're out of position, you may throw the timing off for the entire group. If you're careless about covering your assignment, teammates have to abandon their duties to bail you out.
- Sometimes, of course, you'll need to cover for teammates since everyone needs a little help now and then. But don't poke your nose

into their business or get in their way. Usually you support your teammates best by playing your position to perfection.

6.1.1.1 5-S and Visual Cues Campaigns

Before beginning the identification and execution of Kaizen Improvement Events, the newly formed teams should begin their continuous improvement efforts through workplace environment optimization. Creation of efficient work environments facilitates Lean Vision, or the ability to "see" the waste more clearly. To begin optimizing the work place environment a formal, but short, campaign to implement the 5-Ss and establish Visual Cues should be the first effort of mobilization. Although these are short campaigns (generally one week is sufficient, two weeks at the most), it must be emphasized that their purpose is to make significant progress during the campaign period, but that the improvement efforts of 5-S and Visual Cues is one that continues for the life of the Lean Maintenance Operation.

5-S/Visual Cues Efforts—Referring to Chapter 5, Section 5.1.1.1 of this handbook for a review of 5-S activities, the very first effort should be to remove everything that is not used in job performance from the workplace. In at least 90% of plants you can expect dramatic results from this effort. In fact, one of your first visual motivators may be in the form of before and after photos; they will provide a continuous reminder of how badly cluttered the workplace can become when 5-S isn't practiced. Parts and consumables, other than those for daily and scheduled use, should be returned to stores. Be sure to apply the red tag method for moving the other unneeded items to a central location to ensure that no items needed in another workplace are discarded.

Continue with the 5-S activities by putting everything in order and establishing standard locations using visual indicators. Clean the workplace—inside-out and upside-down, including tools and equipment located within the workplace—and establish a standardized daily routine for keeping everything in order and clean. Use visual cue cards (similar to the signs stating, "Last one out turn off the lights") as reminders for performing orderliness and cleanliness checks, not only at the end of each day or shift, but also following routine daily job completion, such as:

Return all unused parts to stores
Did you return your tools to their proper storage?

Take all used and replaced parts and consumables to industrial waste and any others that apply to your workplace operations.

A contribution that the Lean Maintenance Project Manager can make to aid this campaign is to provide a series of posters illustrating the Tools of Lean to each team. Posting of these tools will serve as a continual reminder of the fundamental concept of Lean—to do away with all non-value-adding activities in their work processes.

Continuous Improvement Efforts—Most of the continuing efforts by the action teams following the 5-S campaign will involve identifying processes, analyzing them and identifying improvements, implementing the improvements and studying the results and taking follow-up actions either to make the change permanent or to begin analyzing them again—the PDSA process of Kaizen events.

While this approach is effective, particularly for Action Teams still new to the improvement process, it is important for the team membership to understand that Lean thinking is all about thinking outside of the box. They should not feel constrained to use this approach and only this approach. Although Kaizen literally translates to continuous improvement, in truth, it means small incremental improvement. It is best done in little pieces, in the slivers of time that arise even in busy shops. This approach better integrates improvement with daily work, engages everybody rather than just a small team and fosters learning by everyday practice. Kaizen does not need to be an event. It does not even need to be perceptible. In practice, employees who have been turned loose for improvement, have mixed daily work and improvement so artfully that only close observation would reveal the small improvements occurring. There is no wrong approach as long as improvements in work practices and processes continue to be made.

6.1.1.2 Autonomous Operator Maintenance

Many of the teams formed at the beginning of Lean Mobilization will have as their workplace a location in the production area. These teams should have a production line operator assigned to their team. Where this is not the case, production line operators should be assigned to teams responsible for maintaining the equipment that they operate. The reason for this is Autonomous Maintenance (AM) or routine maintenance performed by the production line operators. The Maintenance Manager and Production Manager will need to agree on and establish policy for:

- Where in the production processes autonomous maintenance will be performed
- What level and types of maintenance the operators will perform
- How the work process for autonomous maintenance will flow

When all these conditions have been defined, production line operators must receive training for their specific autonomous maintenance activities. Typically tasks such as cleaning, inspection, minor adjustments and lubrication will become the operator's responsibility. Lubrication task written procedures should be adequately detailed showing lube points, types of lubricants and amounts. Proper lock-out/tag-out procedures should be included as part of the operator PM tasks (AM).

A significant amount of formal training may be required to teach operators the skills needed to take on this role. A skills assessment should be performed to identify the types of training required to bring operators to the desired level of performance. This typically requires a combination of formal and on-the-job training. If feasible, rotating operators into a tailored maintenance apprenticeship program is an excellent approach. Another approach is scheduling a maintenance person with an operator for some period of time to provide the on-the-job portion of their training. This serves to get the operator started off in the right direction and help to ease what can sometimes be a difficult culture change.

A very effective team training method oriented towards autonomous maintenance is deployment of One-Point Lessons. One-Point Lessons are short visual presentations on a single point that are presented on one or two pages. They are supported by diagrams, photographs or drawings. By following the following seven-step process, you can quickly create effective and well retained one-point lessons.

Step 1: Learn when and how to use each of the following types:
- Basic knowledge
- Problem case study
- Improvement case study

Step 2: Discover need for one-point lesson
- Identify team weaknesses, skill deficits, relevant equipment problem areas, etc.

Step 3: Launch project
Step 4: Create lesson
Step 5: Implement one-point lesson
Step 6: Archive and share one-point lessons
Step 7: Continue to discover.

Before the Autonomous Maintenance (AM) is turned over to the operator, a final qualification process should be used to certify the operator. A simple combination of a written exam and hands-on skills demonstration should be sufficient.

Production AM Coordinator—In larger manufacturing operations, the scope of Autonomous Maintenance may be sufficiently large to warrant assignment of an Autonomous Maintenance Coordinator. His responsibilities would include:

- Provide Maintenance Planner with up-to-date autonomous maintenance resource data
- Provide liaison between Planning and Scheduling and Production Line Supervision for scheduling and for rapid notification of maintenance opportunities (critical equipment becomes available for outstanding maintenance)
- Provide Work Order processing support within the Production Department
- Provide liaison between Assistant Maintenance Manager/Maintenance Supervisors to ensure there are adequate numbers of trained and qualified autonomous maintenance resources

5-S/After Visual Cues on the Production Line As Autonomous Operator Maintenance is activated, a short 5-S/Visual Cues period should be formally scheduled in those production areas where AM is being performed. The Visual Cues effort here should emphasize the operation and autonomous maintenance aspects of the production equipment in addition to the environmental aspects of the workplace. The application of visible labels and cue cards should include:

- Meters and Gauges—normal operating ranges and high and low red line values
- Autonomous Operator Maintenance
- Lubrication—lubrication points, lubricant type and amount, frequency
- Cleaning—area/item to be cleaned, cleaning agent, frequency and/or criteria for cleaning
- Adjustments (speed, flow, pressure, etc.)—location of adjustment, conditions for adjusting, adjustment value, frequency and/or criteria for performing adjustment
- Inspections—item to inspect, conditions to inspect for, inspection criteria, etc.
- Safety Measures—operating and performing AM
- Others—depending on production equipment type, configuration and autonomous operator maintenance being performed

6.1.1.3 Action Team Leader Knowledge Sharing

Immediately following the 5-S/Visual Cues Campaign the Lean Maintenance Project Manager should establish weekly meetings of the team leaders from each Action Team. For the first two or three meetings the Lean PM should act as facilitator, but after those initial sessions his continued attendance is discouraged unless there are specific issues he must address with the team leaders.

The purpose of the weekly team leader meetings is to (briefly) review each team's accomplishments during the preceding week, concentrating on "lessons learned," specific problems encountered and the solutions that were identified and implemented. Not only do these meetings foster broad, uniform knowledge levels among the Action Teams, but they also serve to sustain continuous improvement. Through a subtle, but nonetheless strong, sense of a need to show progress to the other team leaders, a sustaining force is created among them.

6.1.1.4 Completing Maintenance Mobilization

When empowered, self-directed action teams have been formed, the 5-S/Visual Cues Campaign has been completed and Autonomous Operator Maintenance instituted, completing the mobilization is the final element of Phase 4. The action teams are now fully chartered to identify and eliminate waste—non-value adding activities—within their work areas/work centers. All the tools of Lean Maintenance are at their disposal. The Lean Tools posters provided by the Lean Maintenance PM will serve to keep these tools current in everyone's mind. Additionally, the following principles for performing Kaizen (improvement) activities should be provided to your action teams:

1. Discard conventional fixed ideas.
2. Think of how to do it, not why it cannot be done.
3. Do not make excuses Start by questioning current practices.
4. Do not seek perfection for the future, do it right away even if for only 50% of target.
5. Correct it right away, if you make mistake.
6. Do not spend money for Kaizen, but use your wisdom instead.
7. Wisdom is brought out when faced with hardship.
8. Ask *WHY?* five times and seek root causes.
9. Seek the wisdom of ten people rather than the knowledge of one.
10. Kaizen ideas are infinite.

Leadership again plays a vital role. Without it, the Lean transformation can easily grind to a halt following the 5-S/Visual Cues Campaign. Everyone begins to sense a feeling of completion because getting to this point, particularly with the department-wide conversion to action teams, has been daunting and a continual challenge to leadership. Fight the urge to let up, because you are going to need all of your skills and energy to keep the teams motivated and LEANing.

6.1.2 Mobilization Brings Change

6.1.2.1 New Roles for Management and Supervision

With the formation of action teams—which are empowered, self-directing and team activity oriented—the roles of management and supervision require rather dramatic changes to take place. Instead of directing and controlling, the new role is support.

Empowerment stresses participation and autonomy. Decisions on broad-based issues such as implementation of elements of RCM* or introduction of a new reward system, are made after management has entered into a dialogue with the affected employees. In their new roles, managers provide overall guidance for the work that is clear and engaging. They also offer hands-on coaching and consultation to help employees avoid unnecessary losses of effort, to increase task-relevant knowledge and skills, and to formulate uniquely appropriate performance strategies that generate real process improvements.

Management and Supervision should also be responsive to requests from employees to ensure that the resources required for performance are available when needed. Every complaint should be considered an opportunity for improvement, and people are encouraged to turn their complaints into ideas for improvement. Employee empowerment can degenerate into exploitation if changes at the first level of supervision are not continuously reinforced by changes throughout the management hierarchy. A strong employee voice is needed to ensure that shop floor concerns are heard at all levels of management.

6.1.2.2 A Change of Organizational Focus

One of the dominant characteristics of Lean Organizations as compared to traditional organizations is the flattening of the organization's

---

* "Reliability Centered Maintenance is a process used to determine what must be done to ensure that any physical asset continues to do what its users want it to do in its present operating context."—John Moubray

structural hierarchy—fewer layers of middle-level management. This is a direct result of creating and empowering the self-directed action teams. Along with the flatter structure, a shift in focus needs to occur within the organization if empowerment and continuous improvement are to be sustained. The emphasis now is on recognition of the employee as the plant's most valuable asset.

Reward and Recognition—The desire to attain status in organizational settings is human nature. Attempts to eliminate status through de-layering, or removal of status markers such as an assigned parking space, will find new variations spring up in their place. Instead of working against such human instincts, managers need to recognize and reward employees through status recognition in flexible ways. Promoting employees who teach and help others to team leadership roles reinforces the team culture. It is realized that promotion cannot be used over-liberally to reward exemplary performance, especially in a flatter organizational structure. Therefore, fluid forms of recognition may need to be adopted. These include bonuses, performance awards, certificate of appreciation and one-shot responsibilities such as leadership in a system-commissioning task or a plant-refurbishment project.

A wide variety of compensation programs that take into account factors other than rank, experience and length-of-service are being used in some progressive organizations. In a pay-for-skill program, maintenance tradespersons are paid for acquiring and applying new skills and knowledge required by their jobs. Many manufacturers have implemented similar reward programs to develop their multi-skilled employees. Pay-for-performance and goal-sharing programs award bonuses are linked to group performance. For example, action team excellence is a pay-for-performance program that rewards the performance of the team as a whole. Within each team, rewards are distributed on the basis of factors such as experience. If an organization stresses a team structure, the compensation structure should promote teamwork, not undermine it. The following are critical success factors for a reward and recognition system that encourages teamwork:

- Top management commitment to teamwork and the concept of team-based rewards and recognition
- Management is available and visible
- Employees are regarded as the organization's most valuable assets
- Employees value empowerment and involvement as a form of reward and recognition
- The organization relies on structured processes, policies and documentation

- A strong network is in place for vertical, horizontal, diagonal, intra-team and inter-team communication
- A performance measurement is in place
- Employees participate in training

Offering employees the "right" rewards alone is unlikely to produce sustained empowerment. The power of such methods wears off with use, creating dependency to maintain commitment. Trust, involvement and autonomy are the lasting ingredients that drive human energy and activate the human mind.

Education and Training Empowerment will degenerate into abandonment if employees fail to get the right tools, training on their use and support in their implementation. Educational resources, which can include technical consultation as well as training, must be available and accessible to employees with identified needs. For instance, the specialists of maintenance departments are called upon to upgrade operators to autonomous operator-maintainers. However, the training should not be limited to transfer of technical skills and knowledge needed for optimal task performance. It should also cover generic matters like the business imperatives peculiar to the organization (what determines the value of its product and services to customers), problem-solving techniques, team dynamics and facilitation skills. Additional training for managers should address issues such as the new roles (leader, communicator, coach, resource providers) they fill in the Lean environment, and the new management behaviors that will align efforts and generate commitment for organizational goals.

6.2 EXPANDING THE LEAN MAINTENANCE TRANSFORMATION (Phase 5)

There are two areas influenced significantly by the Lean Maintenance Expansion (Phase 5) and several other areas experiencing a lesser impact, but all are important to successful implementation of Lean Maintenance. Because Phase 5 expansion doesn't directly involve or conflict with Lean Maintenance Pilot or Mobilization efforts, it does not need to wait for the completion of Phases 3 and 4 to begin performing Lean expansion activities. However, in order not to draw any attention away from the pilot Kaizen Events and the 5-S Campaign, expansion activity should be delayed until they are completed (refer to the POA&M for Lean Maintenance Transformation in Section 4.4).

6.2.1 Lean Expansion Major Efforts

Purchasing will have a substantial effort in providing Lean Support to the Maintenance Organization, but much of what they learn during the expansion phase will also be of significant benefit during the plant-wide Lean transformation. In addition to the purchasing operation—actually integrated with purchasing efforts—the Maintenance Engineering group will also have a sizable effort to complete their Lean transformation.

6.2.1.1 Expanding to Purchasing

Taking Lean to the Supply Chain—We have referred to Standardization and JIT, or Just-in-Time, several times as effective tools for implementing Lean Maintenance. Nowhere does it have a more significant effect than in maintenance stores—repair parts and materials. The potential for cost savings can be as much as 30% of the entire maintenance budget. Reactive maintenance organizations typically have a large inventory of spare parts and at the same time engage in excessive emergency purchasing activities to address breakdowns. Additionally they carry literally hundreds of different maintenance materials and consumables such as lubricants, common tools, wipes, paper, forms, pens and pencils (some of these latter items may be issued from an administrative supplies storeroom, but the problem is the same). Standardized buying of these consumable items can save substantial sums. And, as we transitioned toward a proactive TPM culture with more planned and scheduled work, the need for maintenance is identified early enough to be able to order parts and materials and receive them in a JIT scenario, before failure or breakdown occurs.

Standardized Materials/Consumables

One of the first areas that the MRO Storeroom should begin its standardization efforts (because of the probability of the largest savings) is with lubricants. Many OEMs call out specific lubricants by manufacturer as well as by grade and specifications because of affiliations or other favorable agreements. This does not mean that the lubricant of that manufacturer is the only permissible lubricant for that equipment. It does not even mean that lubricants meeting the specification (society) cited is the only acceptable lubricant for that equipment. Just as there are many manufacturers of lubricating products, there is no shortage of organizations generating lubricant specifications (SAE, API, ASTM, etc.).

Most manufacturer's lubricants and most professional society lubricant specifications are duplicated by other manufacturers and specifying societies. In order to standardize your storeroom selections a database of lubricant equivalents needs to be generated and then used to identify single—or maybe two or three—source suppliers. Many companies and even professional (specifying) societies have developed these equivalency databases and are more than willing to provide them to you for a nominal fee. The savings you realize from standardization will certainly offset the small price you will pay for the information. Apply similar methods to standardize your storeroom's other multiple, common-use items.

Standardized Parts and Suppliers

The improved organization of the spare parts storage locations during changeover to TPM helped to eliminate duplication of inventory and facilitated calculation of appropriate minimum and maximum stocking levels and economic order quantities. There are several tools or practices that can be implemented by purchasing and between purchasing and the MRO Storeroom that can reduce inventories and at the same time provide improved parts and materials support for the maintenance function.

A special note: a key element for many of these practices to be effective is having complete and accurate equipment and repair part information in the CMMS database. Many of these techniques require contracts for implementation. Purchasing personnel should receive training in contract types and in contract negotiation if they are to be effective in applying these practices.

1. Develop planning and forecasting techniques to stabilize the purchasing and storeroom management process. This method requires that a long-term equipment plan is developed. Equipment Bills of Material are entered into the CMMS system, as soon as the purchase order for new equipment is issued. (Some are as much as two to three years out.) That way component parts to support the new equipment are entered into the system in preparation for future purchases. In addition, insurance spares and components for insurance spares are identified early in the process, so that plans can be put into place to store these materials. By knowing what you are buying, and establishing insurance spares and parts identification, you will minimize future stock outs. This leads to improved requirements planning for storeroom materials. Through failure analysis studies (manufacturer knowledge), previous experiences with the parts, and through Delphi methods (group discussion among maintenance and engineering), parts planning can take place more effectively to determine its replenishment cycle. This method enhances planning and stabilizes the buying process.

2. Develop consignment inventories, where the manufacturer/vendor sends material to the storeroom. Payment for the material occurs when the stock is "issued to a work order." This minimizes stock outs, as the manufacturer/vendor is allowed to send material at peak times, since it is not paid for until used. This strategy allows the manufacturer to produce by convenience and schedule, rather than by fluctuating demand. This, in turn, benefits the facility as well, as it reduces inventory carrying costs, reaffirms Lean inventory management and increases cycle time as downtime is minimized with this method.

3. Establish vendor-managed inventories (VMI). A vendor-managed inventory is a method in which the vendor is given the responsibility to maintain good inventory practices in replenishment, in ordering and issuing the materials. The purchase order is issued based on the vendor's recommendations. The vendor is charged with the responsibility of controlling costs, inventory levels, the sharing of information with the facility and improvements in the process. The vendor is evaluated in much the same manner as a member of your staff is in achieving the goals required by the facility. While few companies employ the strategy of VMI, the success rate of its incorporation has been noteworthy. A successful implementation of this strategy takes detailed planning, solid management direction and commitment and an organization structure to allow the VMI to function as a member of the staff.

4. Develop a vendor rating system that objectively rates your vendors quantitatively and qualitatively. Quantifications include delivery days and price variances. Qualitative measurements are based on pricing, quality of materials received, delivery and service. Vendors are then certified (vendor that supplies materials on time, with the right quality, at the most competitive price and with the best service), targeted for improvement or eliminated. The goal is to reduce the number of vendors and consolidate purchases. This ultimately affects volume discounting, quality improvements and more competitive pricing.

5. Establish a vendor partnership contract agreement. This goes beyond current blanket orders and standard written agreements. It is a written agreement where the plant and the vendor share information in a mutually confidential manner. All aspects of the business pertaining to the vendor's role in the organization are discussed. Future plans and developments, as well as current equipment concerns, are discussed in scheduled meetings. A vendor is considered a part of the staff, and as such is given responsibilities to

reduce costs, improve profits, improve cycle time, reduce failures due to equipment breakdowns, improve quality and assist the facility in improving its competitiveness. A written contract binds the vendor and the plant for past, current and future considerations. It is a method used to stabilize pricing, reduce inventories and keep all stock defect free.

6. Utilize World Class Logistics Support. Small package deliveries are timed to deliver products on a just-in-time schedule. No deliveries are unscheduled. LTL carriers are contracted not only for a scheduled time between point-of-pick-up to point-of-delivery, but also for pricing and claim settlements. Planned Deliveries are now considered a normal mode of operation for the World Class storeroom. Emergency deliveries are pre-planned, with pricing already in effect for such a conveyance. The emphasis on logistics is to reduce the number of carriers delivering products to the facility, reduce the length of carrier and package delivery time to the facility and optimize response time for maintenance to receive a part and replace it in equipment. The carriers that are a part of the logistics team are also considered to be a part of the staff. Information is shared with them concerning future deliveries, current issues and problems and cost improvements. It is a win/win situation when carriers and small package services are included in planning activities and scheduled meetings.

6.2.1.2 Expansion to Maintenance Engineering

Predictive Maintenance and Optimizing Maintenance
Although we have addressed Maintenance Engineering's role in the Lean Maintenance Transformation as it relates to Action Teams and their activities, an area not addressed is the role of Maintenance Engineering in optimizing maintenance. One of the most widely used tools in this regard is Predictive Maintenance (PdM) to forecast necessary maintenance actions. Depending on the quantity and kinds of production equipment in your plant, the array of PdM techniques can range from as few as two or three to as many as ten or even more. (see Chapter 3, Section 3.1.6)

Because of the highly specialized skills involved in operating much of the equipment used in PdM as well as in interpreting the results, and the requirement, in many cases, of specialized and costly laboratory and/or computing facilities to perform analysis, the question of outsourcing PdM services as an economy must be answered. Analyzing the alternatives and contracting for those PdM activities that you have decided to outsource are accomplished during expansion.

In the past, when the benefits of in-house capabilities were emphasized in management thinking, internal resources typically performed nearly all predictive maintenance activities. External suppliers were used only under the following situations:

- The in-house Maintenance Engineering group did not have sufficient capacity to meet peak demand; in such cases, short-term outsourcing would be used to fill the shortfall
- The expected volume of a particular PdM application was too small and the variety of maintenance-related specialist skills too wide to justify an in-house specialist
- The organization did not have the expertise and specialized facilities to perform the PdM work; the cost of developing such capabilities and assets in-house would be prohibitive while there were established suppliers in the market to provide the required services

In Lean Thinking, a new trend has emerged that subscribes to the concept that unprecedented business performance can be achieved if the skills and resources are leveraged to focus on a set of core competencies—a bundle of skills and technologies that enables an organization to provide a value added benefit to customers. Therefore, predictive maintenance activities for which the company has neither a strategic need nor a special capability are prime candidates to be outsourced. The PdM services typically outsourced include those requiring costly training and certification, substantial space requirements (labs, etc.) and expensive testing equipment with a low usage factor.

The selection of predictive maintenance service-delivery options should not be regarded as a purely tactical matter. The decision should be made in the context of your plant's overall business strategy. When companies consider outsourcing of their predictive maintenance activities as a strategic option, they need to answer three key questions:

1. What should not be outsourced?
2. What type of relationship with the external service supplier should be adopted?
3. How should the risks of outsourcing be managed?

What should not be outsourced?—There are two key strategic issues that determine the choice between outsourced and in-house provided services.

The first factor is the potential for achieving a competitive edge by performing the work internally. If management perceives that excellence in performing certain predictive maintenance services—done cheaper, better or more timely—will enhance the company's competitiveness, such services should be carried out internally. The second factor is the degree of strategic vulnerability if the work is outsourced. If there is insufficient depth in the market, an overly powerful supplier can hold the company hostage. On the other hand, if the individual suppliers are too weak, they may not be able to supply quality and innovative services as well as the buyer could by performing the work internally.

Knowledge is another important dimension that affects vulnerability. It is extremely risky to outsource work when the company does not have the competence either to assess or monitor suppliers, or when it lacks the expertise to negotiate a sound contract. The caveat that companies should not outsource those activities that are crucial elements of their core competencies is often not heeded when outsourcing decisions are driven by cost-cutting and headcount-reduction criteria. As a result, control of activities critical to establishing the company's competitive advantage can be unwittingly ceded to suppliers. Another common fallacy in making outsourcing decisions is to regard core competencies as things that we do best. This misconception is damaging, as it encourages management to outsource activities with which it is having problems. If the company has difficulty in managing an internal supplier, it probably cannot communicate its requirements adequately to the external supplier. Thus, internal problems are traded for new problems of dealing with external suppliers. It will be even more devastating if the problematic activity over which the company relinquishes control is a critical link in its current or future value-creation process. When an external supplier offers a significant cost-saving deal on the company's core activities, management should refrain from outsourcing them. Instead, the internal service provider should be challenged to improve its cost effectiveness, using the supplier's offer as a benchmark of performance.

When a predictive maintenance service, which can be one of the things that we do best, has been classified as a non-core activity, it can be considered for outsourcing. However, the decision now depends on the relative costs of in-house and external provision of that service. Apart from the direct costs involved, the relevant transaction costs to consider are:

- In-house provision: continuing R&D, personnel development and infrastructure investment that at least match those of the best supplier to maintain a competitive edge; overhead for managing the in-house activities.
- Outsourcing: the costs of searching, contracting and controlling the outsourced activities.

If it is found to be more cost effective to keep an exemplary but non-core capability, the company should explore the possibility of commercializing the expertise to serve the needs of noncompetitors. For example, if you have equipment and expertise for performing vibration monitoring and analysis but only exercise it for 20% of the time available to exercise it, market the service to companies in noncompetitive markets—you can afford to under-price other vibration monitoring and analysis providers, because the equipment and expertise you possess isn't producing a return for 80% of the available time (exclusive of the "other" work responsibilities that the expert has during that time).

6.2.1.3 Expansion to IT Department

The Information Technology group should be actively pursuing continuous improvement of the support efforts that they provide for the maintenance operation. Upgrades of CMMS, periodic CMMS refresher training and new personnel training, frequent communication with CMMS data input personnel and data (reports) output users are ways to maintain optimized CMMS effectiveness and utilization.

Additionally, because the technology associated with IT hardware and software is such a rapidly evolving field, the IT group should also be evaluating new products and applications. Through frequent communications as well as an understanding of the maintenance operation, the IT group should determine suitability and efficacy of new technology for maintenance and maintenance support operations.

At the device level, RF systems, wireless technology and palm computing top the list of technologies to be considered. Wireless technology offers the promise of implementation of e-maintenance. It provides low-cost, short-range radio links between mobile computing devices, mobile phones and other portable devices. Integration of wireless technology and palm computing into the maintenance-execution process promises accessibility and mobility. Enterprise systems are accessible from remote sites and controlled documents such as manuals, multimedia work instructions, safety information and inspection plans can be downloaded for field use on location.

There is a gradual trend to shift from CMMS to Enterprise Asset Management (EAM) systems. EAM is a global information management system for corporate-wide, multi-site application. As their ease of use and utility improves we can expect this trend to grow. The IT group should stay abreast of this kind of application technology to determine when it, or another application, is right for their company.

Some Technology Derived Potential Benefits—Improved accuracy and consistency. Diagnosis, maintenance work performed and parts replaced are documented on the spot through structured responses to work steps displayed on the palm top. Time and costs are recorded against correct job and location. Results of inspection and audits can be recorded in real time using standardized responses to ensure consistency of captured data. The non value-added and error-prone operation of transferring operator logs to computer systems can be eliminated.

Stores management—Goods are checked out from stores against a work order or a location and the transaction is recorded in real time. The enterprise system is notified as soon as any withdrawal takes place.

On-site permits—Lockouts and isolation can be performed and recorded on location. Equipment tagged with bar codes prevents possibility of error.

7

Sustaining Lean— Long-Term Execution

7.1 SUSTAINING CONTINUOUS IMPROVEMENT (PHASE 6)

Leadership!

Which tool is the most effective in sustaining the Lean Maintenance Transformation? Of course every one of them is important, but the display of leadership and commitment will be the attributes that determine the long-term success of the Lean Enterprise.

At the Corporate Executive level, Plant Management level, Department Management level, Lean PM level, Supervisor level and Action Team level, leadership qualities are the fuel that will keep you moving at highway speeds or force you to drive on the highway shoulder as you coast to a stop when you run out.

In the application of leadership, it is important to understand that the quality of leadership is something experienced by those being led—and it is evaluated by them. At the corporate level you find the winning leaders are those who perceive their organizations as creations that can be recreated any way they wish to make up. Effective leaders are in touch with the way things work and are constantly challenging people to generate ideas that will make things work better. They discourage positional debates, except as learning devices. They constantly encourage the movement to better ways that can be reviewed in the future. Leaders are not concerned about the status of people with improvement ideas and energy. An effective leader knows that the reality of the person is not in his title, position, or name, but in what his actions reveal and accomplish. Commitment!

From the corporate executive level down, commitment to the Lean Transformation and to long-term Lean Thinking must be firm, it must be communicated and it must be visible. Demonstration of this entails:

- Top management active commitment to teamwork and the concept of team-based rewards and recognition
- Management is available, involved and visible
- Employees are regarded as the organization's most valuable assets
- Employees value empowerment and involvement as a form of reward and recognition
- The organization relies on structured processes, policies and documentation
- A strong network is in place for vertical, horizontal, diagonal, intra-team and inter-team communication
- A performance measurement system is in place
- Employees participate in training definition and development

7.1.1 Applying the Tools

Continuous Improvement!

As initial improvements are made, strong reinforcement from management should be provided through recognition, rewards (as appropriate), encouragement towards ever-higher achievement and development of action plans and goals for future (short-term) completion. Managers and supervisors need to spend the majority of their time with the action teams, actually participating with them in their improvement activities through suggestion, opinion, concurrence and general encouragement to achieve more. After several months and several process improvement projects, the methodology of continuous improvement will become ingrained and accepted by the team as their normal way of doing business. As this occurs, it is not a signal to "leave them on-their-own." A sudden decrease in manager and supervisor visibility and participation will always be interpreted as indifference and loss of interest resulting in the team's loss of interest as well.

Continuous Improvement involves application, as appropriate, of all the tools available for adding value and eliminating waste—action teams should not feel limited to the Lean Transformation tools that we have discussed to this point. Any thought process and any practices that yield improved ways of doing things are not only appropriate, they are strongly encouraged.

Innovation is one of the keys that unlocks the future!

7.1.1.1 Optimizing Maintenance Using Lean Tools

Maintenance Engineering

In addition to the Action Team activities of continuous improvement, Maintenance Engineering has a major, well-defined role in TPM optimization. Using the results from Predictive Maintenance (PdM) and Condition Monitoring (CM) and Preventive Maintenance task analysis, the job of Maintenance Engineering is to eliminate unnecessary maintenance activity. The responsibilities of Maintenance Engineering for the establishment and execution of maintenance optimization using CMMS Unscheduled and Emergency reports, Planned/Preventive Maintenance reports and CM/PdM analysis include:

- Develop CM tests and PdM techniques that establish operating parameters relating to equipment performance and condition and gather data
- Analyze CMMS reports of completed CM/PdM/Corrective work orders to determine high-cost areas
- Establish methodology for CMMS trending and analysis of all maintenance data to make recommendations for:
  - Changes to PM/CM/PdM frequencies
  - Changes to Corrective Maintenance criteria
  - Changes to Overhaul criteria/frequency
  - Addition/deletion of PM/CM/PdM routines
  - Establish assessment process to fine-tune the program
  - Establish performance standards for each piece of equipment
  - Adjust test and inspection frequencies based on equipment operating (history) experience
  - Optimize test and inspection methods and introduce effective advanced test and inspection methods
  - Conduct a periodic review of equipment on the CM/PdM program and eliminate that equipment no longer requiring CM/PdM
  - Remove from, or add to, the CM/PdM program equipment and other items as deemed appropriate
- Communicate problems and possible solutions to involved personnel
- Control the direction and cost of the CM/PdM program

Defining Optimum Maintenance Frequencies—A common concern in developing and refining TPM programs is the time duration equipment should be operated between overhauls.

Consider a bank of compressors with historical experience as illustrated in Table 7-1.

Table 7-1
Single Component Failure

Compressor Number	Run Hours Between Failures
8	660
5	780
1	520
3	720
6	570
2	480
8	675
4	715
5	560
7	700
6	645
3	545
7	650
3	625
5	590
6	590
8	620
2	750
4	550
1	640
7	585
4	600
2	640
1	710
Average	630

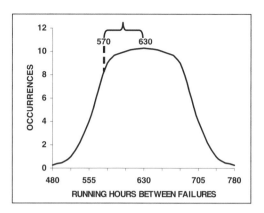

Figure 7-1 Time Based Determination of Overhaul Frequency

Most failures of single components follow the normal law: If the point in time is desired where 50% of the compressors fail, 630 hours would be chosen. If the goal is to prevent 70% of the failures, each compressor would be taken out of service after 570 hours of operation (see Figure 7-1).

In maintenance operations, interest is usually focused on groups of parts or machines that are, in effect, systems because of the interrelation of the parts. They need to be analyzed in the following manner in Table 7-2:

Table 7-2
Statistical Determination of Overhaul Frequency

Part	No. Used	Failure Rate Per 5,000 Hrs.	Total Failures Per 5,000 Hours.
Casing	1	2.1	2.10
Shaft	1	8.4	8.40
Bearing—Type A	4	12.2	3.05
Bearing—Type B	7	16.4	2.34
Bearing—Type C	5	8.2	1.64
Valves	16	5.5	.33
Valve Springs	16	18.6	1.16
Cylinders	8	.5	.06
Pistons	8	1.2	.15
Connector Rods	8	.2	.02
Pins	16	6.1	4.00
Rings	24	20.8	.86
Gaskets	4	18.7	4.67
Total failure per 5,000 hours			25.18

Therefore, estimated time between failures (ETBF) for a system as a whole is:

$$\text{ETBF} = \frac{\text{Time Increment}}{\text{Total Failures During Time}}$$

$$= \frac{5,000 \text{ Hours}}{25.18 \text{ Failures per 5,000 Hours}}$$

Mean Time Between Failures = 198.57 hours or 200 hours

For random failures, the optimal Condition Monitoring frequency is given by the following formula:

$$n = \frac{\ln\left(\dfrac{\left(\dfrac{-\text{MTBF}}{T}\right)\text{Ci}}{(\text{Cnpm} - \text{Cpf})\ln(1-S)}\right)}{\ln(1-S)}$$

Where:

- T = Age (time) between the point at which the failure can be first detected and the actual failure—also known as PF interval
- n = Number of inspections during the PF interval T
- MTBF = Mean Time Between Failure
- Ci = Cost of one inspection task
- Cpf = Costs of correcting one potential failure
- Cnpm = Cost of not doing preventive maintenance inclusive of the operational costs of failure
- S = Probability of detecting the failure in one inspection (task effectiveness)

The frequency of predictive maintenance tasks has nothing to do with the frequency of failure and nothing to do with the criticality of the item. The frequency of any form of condition-based maintenance is based on the fact that most failures do not occur instantaneously, and that it is often possible to detect the fact that the failure is occurring during the final stages of deterioration.

Figure 7-2 shows this general process. It is called the *P-F curve* because it shows how a failure starts and deteriorates to the point at which it can be detected (the potential failure point *P*). Thereafter, if it is not detected and suitable action taken, it continues to deteriorate—usually at an accelerating rate—until it reaches the point of functional failure *F*.

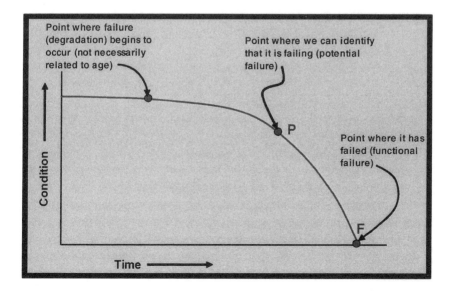

Figure 7-2 P-F Curve

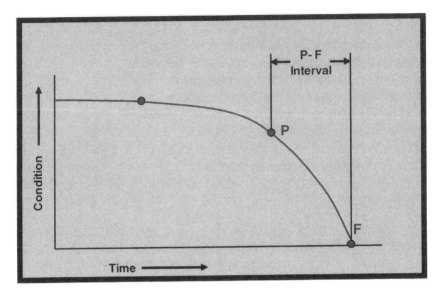

Figure 7-3 P-F Interval

The amount of time (or the number of stress cycles) which elapse between the point where a potential failure occurs and the point where it deteriorates into a functional failure is known as the *P-F interval*, as shown in the Figure 7-3.

The P-F interval governs the frequency at which the predictive task must be done. The inspection interval must be significantly less than the P-F interval if we wish to detect the potential failure before it becomes a functional failure.

The P-F interval can be measured in any units relating to exposure to stress (running time, units of output, stop-start cycles, etc), but it is most often measured in terms of elapsed time. The amount of time needed to respond to any potential failures that are discovered also influences condition-based task intervals. In general, these responses consist of any or all of the following:

- Take action to avoid the consequences of the failure
- Plan corrective action so that it can be done without disrupting production and/or other maintenance activities
- Organize the resources needed to rectify the failure

The amount of time needed for these responses also varies, from a matter of hours (say until the end of an operating cycle or the end of a shift), minutes (to clear people from a building which is falling down) or even

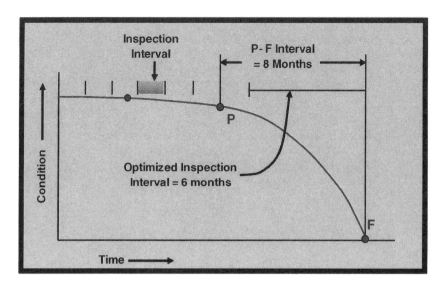

Figure 7-4 P-F interval of 8 months and a checking interval of 1 month yield an optimized inspection interval of 6 months

seconds (to shut down a machine or process that is running out of control) to weeks or even months (say until a major shutdown).

Unless there is a good reason to do otherwise, it is usually sufficient to select a checking interval equal to half the P-F interval. This ensures that the task will detect the potential failure before the functional failure occurs, while providing an inspection interval of at least half the P-F interval to do something about it. However, it is sometimes necessary to select a checking interval, which is some other fraction of the P-F interval. For instance, Figure 7-4 shows how a P-F interval of 8 months and a checking interval of 1 month yield an optimized inspection interval of 6 months.

If the P-F interval is too short for it to be practical to check for the potential failure, or if the net P-F interval is too short for any sensible action to be taken once a potential failure is discovered, then the condition-based task is not appropriate for the failure mode under consideration.

The inherent problem with time based determinations is that equipment components don't always behave in the same manner during a given time period. Specifically, the overhaul frequency selected from the time based determination above is for a period in the past. Will the equipment perform better in the future because it has been run-in, or will it behave worse because of increasing age? Statistical determinations have much improved predictability, especially as more and more data is added to the statistical database, but still have the inherent problem in expectation of the equipment to behave in the future just as it has in the past.

Condition Based determination of overhaul performance, or of any PM performance eliminates the inherent problem by providing real-time information about the condition of the equipment. If a minimum performance criteria is established for a monitored parameter, then a trend-line of that parameter can predict, with considerable accuracy, when that parameter will exceed the performance cut-off so that preventive or corrective action can be taken just-in-time.

In its simplest form, condition monitoring can employ procedural documentation that provides an area for recording readings of installed meters and gauges when they are pertinent to the preventive maintenance task being performed. Such data can provide a simple means for evaluating the effectiveness of the maintenance and optimizing scheduling. Take for example the scenario depicted in Figure 7-5.

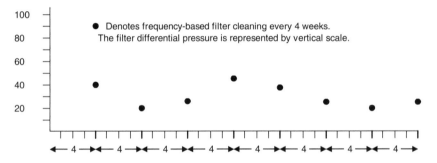

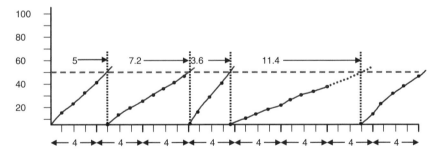

Figure 7-5 Evaluating the Effectiveness of the Maintenance and Optimizing Scheduling

Root Cause Failure Analysis (RCFA)—One of the most important functions of the Maintenance Engineering group is RCFA. Failures are seldom planned for and usually surprise both maintenance and production personnel. And they nearly always result in lost production. Finding the root cause of a failure provides you with a solvable problem removing the mystery of why equipment failed. Once the root cause is identified, a "fix" can be developed and implemented.

There are five basic phases to RCFA:

Phase I: Data Collection—It is important to begin the data collection phase of root cause analysis immediately following the occurrence identification to ensure that data are not lost. (Without compromising safety or recovery, data should be collected even during an occurrence.) The information that should be collected consists of conditions before, during and after the occurrence; personnel involvement (including actions taken); environmental factors and other information having relevance to the occurrence.

Phase II: Assessment—Any root cause analysis method may be used that includes the following steps:
- Identify the problem.
- Determine the significance of the problem.
- Identify the causes (conditions or actions) immediately preceding and surrounding the problem.
- Identify the reasons why the causes in the preceding step existed, working back to the root cause (the fundamental reason that, if corrected, will prevent recurrence of these and similar occurrences throughout the facility); this root cause is the stopping point in the assessment phase.

Phase III: Corrective Actions—Implementing effective corrective actions for each cause reduces the probability that a problem will recur and improves reliability and safety.

Phase IV: Inform—Entering the report on the appropriate RCFA worksheet or report form is part of the inform process. Also included is discussing and explaining the results of the analysis, including corrective actions, with management and personnel involved in the occurrence.

Phase V: Follow-up—Follow-up includes determining if corrective action has been effective in resolving problems. An effectiveness review is essential to ensure that corrective actions have been implemented and are preventing recurrence.

There are many methods available for performing RCFA such as the Ishikawa, or Fishbone, diagram discussed previously. Selecting the right method for RCFA can speed the entire process up so that you can proceed to the "fix" stage more quickly. Brief descriptions of other common methods for performing RCFA are provided here.

Events and Causal Factor Analysis

Events and Causal Factor Analysis is used for multi-faceted problems or long, complex, causal factor chains. While the approach to analysis is similar to that used in constructing a Fishbone Diagram the resulting chart in this technique is a cause-and-effect diagram that describes the time sequence of a series of tasks and/or actions and the surrounding conditions leading to a failure event. The event line is a time sequence of actions or happenings while the conditions are anything that shapes the outcome and ranges from physical conditions (such as an open valve or noise) to attitude or safety culture. The events and conditions as given on the chart describe a causal factor chain. The direct, root and contributing cause relationships in the causal factor chain are shown in Figure 7-6.

Change Analysis

Change Analysis is used for a single failure and free activities associated with the failure to determine how they contributed.

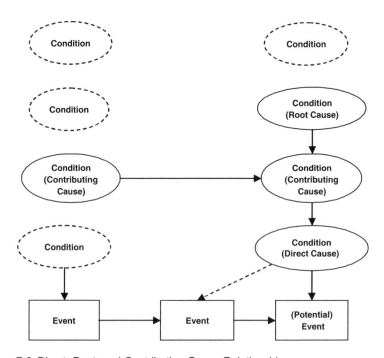

Figure 7-6 Direct, Root, and Contributing Cause Relationships

Change Analysis looks at a problem by analyzing the deviation between what is expected and what actually happened. The evaluator essentially asks what differences occurred to make the outcome of this task or activity different from all the other times this task or activity was successfully completed.

This technique consists of asking the questions: *What? When? Where? Who? How?* Answering these questions should provide direction toward answering the root cause determination question: *Why?* Primary and secondary questions included within each category will provide the prompting necessary to thoroughly answer the overall question. Some of the questions will not be applicable to any given condition. Some amount of redundancy exists in the questions to ensure that all items are addressed. Several key elements include the following:

- Consider the event containing the undesirable consequences
- Consider a comparable activity that did not have the undesirable consequences
- Compare the condition containing the undesirable consequences with the reference activity

Set down all known differences whether they appear to be relevant or not. Analyze the differences for their effects in producing the undesirable consequences. This must be done with careful attention to detail, ensuring that obscure and indirect relationships are identified (e.g., a change in color or finish may change the heat transfer parameters and consequently affect system temperature). Integrate information into the investigative process relevant to the causes of, or the contributors to, the undesirable consequences.

Change Analysis is a good technique to use whenever the causes of the condition are obscure, you do not know where to start or you suspect a change may have contributed to the condition. Not recognizing the compounding of change (e.g., a change made five years previously combined with a change made recently) is a potential shortcoming of Change Analysis. Not recognizing the introduction of gradual change as compared to immediate change also is possible. This technique may be adequate to determine the root cause of a relatively simple condition. In general, though, it is not thorough enough to determine all the causes of more complex conditions. Figure 7-7 shows the six steps involved in Change Analysis.

Appendix A contains a Change Analysis Worksheet together with questions that help to identify information required on the worksheet.

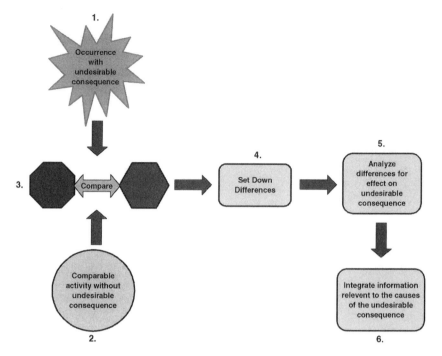

Figure 7-7 Six Steps Involved in Change Analysis

Barrier Analysis

Barrier Analysis is used when the problem is obscure. It is a systematic process that is generally focused on elements that have changed. It compares the previous trouble to identify differences. These differences are subsequently evaluated for the failure. Barrier Analysis is a systematic process that can be used to identify physical, administrative and procedural barriers or controls that should have prevented the failure. This technique should be used to determine why these barriers or controls failed and what is needed to prevent recurrence.

Management Oversight and Risk Tree (MORT)

MORT/Mini-MORT is used to prevent oversight in the identification of causal factors. It lists on the left side of the tree specific factors relating to the occurrence, and on the right side of the tree it lists the management deficiencies that permit specific factors to exist. The management factors all support each of the specific barrier/control factors. Included is a set of questions to be asked for each of the factors on the tree. As such, it is useful in preventing oversight and ensuring that all potential causal factors are considered. It is especially useful when there is a shortage of experts to ask the right questions. However, because each of the management factors may apply to the specific barrier/control factors, the direct linkage or relationship is not shown but is left up to the analyst. For this reason, Events and Causal Factor Analysis and MORT should be used together for serious failures: one to show the relationship, the other to prevent oversight. A number of condensed versions of MORT, called Mini-MORT, have been produced. For a major failure justifying a comprehensive investigation, a full MORT analysis could be performed while Mini-MORT would be used for most other failures. See Appendix A—Management Oversight and Risk Tree for an example of a diagram used in this technique.

Human Performance Evaluation

Human Performance Evaluation is used to identify factors that influence task performance. It is most frequently used for man-machine interface studies. Its focus is on operability and work environment, rather than on training operators to compensate for bad conditions. Also, human performance evaluation may be used for most failures since many conditions and situations leading to a failure ultimately result from some task performance problem such as planning, scheduling, task assignment analysis, maintenance and inspections. Training in ergonomics and human factors is needed to perform adequate human performance evaluations, especially in man-machine interface situations.

Kepner-Tregoe Problem Solving and Decision Making

Kepner-Tregoe is used when a comprehensive analysis is needed for all phases of the failure investigation process. Its strength lies in providing an efficient, systematic framework for gathering, organizing and evaluating information and consists of four basic steps:

a. Situation appraisal to identify concerns, set priorities and plan the next steps.
b. Problem analysis to precisely describe the problem, identify and evaluate the causes and confirm the true cause (this step is similar to change analysis).
c. Decision analysis to clarify purpose, evaluate alternatives, and assess the risks of each option and to make a final decision.
d. Potential problem analysis to identify safety degradation that might be introduced by the corrective action, identify the likely causes of those problems, take preventive action and plan contingent action.

this final step provides assurance that the safety of no other system is degraded by changes introduced by proposed corrective actions.

These four steps cover all phases of the failure investigation process and thus, Kepner-Tregoe can be used for more than causal factor analysis. Separate worksheets (provided by Kepner-Tregoe) provide a specific focus on each of the four basic steps and consist of step-by-step procedures to aid in the analyses. This systems approach prevents overlooking any aspect of concern. Formal Kepner-Tregoe training is needed for those using this method.

Appendix A—Summary of Root Cause Failure Analysis Methods provides a table summarizing each method, when to use it and the advantages and disadvantages of each.

Occurrence causes are usually grouped within seven categories. These categories are listed here together with the most common causes within each category.

1. Equipment/Material Problem
   - Defective or failed part
   - Defective or failed material
   - Defective weld, braze or soldered joint
   - Error by manufacturer in shipping or marking
   - Electrical or instrument noise
   - Contamination
2. Procedure Problem
   - Defective or inadequate procedure
   - Lack of procedure
3. Personnel Error
   - Inadequate work environment
   - Inattention to detail
   - Violation of requirement or procedure
   - Verbal communication problem
   - Other human error
4. Design Problem
   - Inadequate man-machine interface
   - Inadequate or defective design
   - Error in equipment or material selection
   - Drawing, specification or data errors
5. Training Deficiency
   - No training provided
   - Insufficient practice or hands-on experience
   - Inadequate content

- Insufficient refresher training
- Inadequate presentation or materials
6. Management Problem
    - Inadequate administrative control
    - Work organization/planning deficiency
    - Inadequate supervision
    - Improper resource allocation
    - Policy not adequately defined, disseminated or enforced
    - Other management problem
7. External Phenomenon
    - Weather or ambient condition
    - Power failure or transient
    - External fire or explosion
    - Theft, tampering, sabotage or vandalism

7.1.1.2 The Sustaining Environment and Activities

Following the mobilization and expansion of the Lean Transformation, the Maintenance and Maintenance Support Organization should come to a steady-state level of operating in a Lean environment that is characterized by the conditions listed below.

Maintenance Operating Environment

- Effectively Operating Total Productive Maintenance (TPM) Program
    - Zero breakdown objective
    - Safe work practices/workplace safety of paramount importance
    - Eliminates unnecessary maintenance activities
    - Employs standardized work flows and work practices
- Mutually Supporting Attitudes and Practices Between Production and Maintenance
- Autonomous Operator Maintenance Being Performed
- Optimized CMMS—Accurate Inputs and Effective Outputs (Performance Measures and Tailored Reports)
- Organization Characterized by Empowered, Self-Directed Action Teams
- Well Defined, Standardized and Documented Work Procedures/ Processes
- Planning and Scheduling of Maintenance is Optimized
- Visual Work Place Environment
- Maintenance Stores/MRO Storeroom
    - Reduced/minimized inventory levels
    - Zero stockout objective

- Standardized practices for issuing, staging, ordering/re-ordering and reporting (to CMMS) parts and material
- Maintenance Management
  - Role change from directing and controlling to support
  - Spends more time on the shop floor
  - Establishes maintenance performance goals and uses performance measures (CMMS-generated reports) to gauge progress—publishes performance levels
  - Establishes training and qualification program featuring employee participation and multi-skill training

Maintenance and Maintenance Support Activities

- Empowered, self-directed action teams are pursuing continuous improvement using
  - Process Mapping and Value Stream Mapping
  - Visual Controls
  - Mistake/error proofing
  - Value-focused thinking
  - Implement quality at the source by:
    - Identifying (to management and to CMMS) maintenance task procedural documentation errors, corrections and improvements
    - Inspection/verification of tool/part/material specification, type, size, calibration status, etc. when performing maintenance task procedures (Standardized Work Practices)
    - Completely and accurately filling out completed and uncompleted work orders including clear descriptive comments when appropriate
  - Practice Improving Prior Improvements (IPI) by re-analyzing previously completed process improvements
  - Action Team Leaders Meetings weekly to share knowledge and experience
- Maintenance Engineering
  - Analyzes equipment maintenance tasks, failures and operating conditions to optimize maintenance tasks, maintenance effectiveness and frequencies
  - Performs RCFA for all equipment failures/breakdowns—develops equipment maintainability and reliability improvement design changes
  - Performs failed part analysis
  - Through reliability analysis, identifies economical levels and techniques of Predictive Maintenance and Preventive Maintenance and applies results to optimize maintenance activities and equipment reliability

WASTE

CMMS
IMPLEMENTED &
OPTIMIZED

PLANNING &
SCHEDULING

PERFORMANCE
MEASURES

MAINT. ENG.
ANALYSIS OF
PM/CM/PdM & RCFA

EMPOWERED
ACTION TEAMS

AUTONOMOUS
MAINTENANCE

LEAN MRO
STOREROOM & JIT
SUPPLIERS

CONTINUOUS
IMPROVEMENT

AIM YOUR BIG GUNS
AT WASTE

Figure 7-8 The Next Step

- Assesses maintenance organization skill requirements and skill levels to evaluate training and qualification program effectiveness and develop changes as required
- Communicating with IT Group to evaluate and/or improve automated data acquisition, data analysis, equipment history accumulation and generation of tailored reports
- Purchasing
  - Negotiating new supplier contracts to optimize JIT delivery, standardized parts and material procurement and standardized suppliers
  - Continually evaluating purchasing practices to eliminate waste and non-value adding activities
  - Communicating with IT Group to evaluate and/or improve information management practices
- Information Technology (IT) Group
  - Utilizes established communications with maintenance and maintenance support activities to identify new or expanded information management requirements
  - Evaluates new technologies (software and hardware) for application to maintenance and maintenance support activities
  - Provides CMMS user training (refresher and new personnel) to maintenance and maintenance support activities

Appendix A

Checklists and Forms

Maintenance Operation Checklist	
Is the inventory of skills to support the LEAN program available?	
Is training planned to fill skill and technical shortcomings?	
Does the training support the development of predictive analytical skills?	
Does the training support LEAN management and supervisory skills?	
Are the documentation, procedures and work practices capable of supporting Lean Maintenance?	
Are the responsibilities for systems and equipment defined and assigned?	
Are the maintenance history data and results distributed to proper users?	
Is there a feedback system in place for continuous maintenance program improvement?	
Is root-cause failure analysis in use and effective?	
Are failed components subject to post-failure examination and results recorded?	
Are predictive forecasts tracked and methods modified based on experience?	
Are PM task and CM monitoring periodicities adjusted based on experience?	
Does the CMMS fully support the maintenance program?	
Are maintenance cost, cost avoidance and cost savings data collected, analyzed and disseminated?	
Is baseline condition and performance data updated following major repair or replacement of equipment?	
Are appropriate measures of maintenance performance (metrics) in use?	

Figure A-1 Overall Maintenance Operation: Checklist

Table A-1
Recommended Predictive Technology Application by
Equipment Type:

Equipment Item	Recommended Predictive Technologies	Optional Predictive Technologies
Batteries	Battery Impedance Test	Infrared Thermography
Boilers	Hydrostatic Test Airborne Ultrasonic Test Thermodynamic Performance Tests	Infrared Thermography
Breakers	Contact Resistance Test Insulation Resistance Test	Airborne Ultrasonic Test Power Factor Test Insulation Oil Test High Voltage Test Breaker Timing Test Infrared Thermography
Cables	Insulation Resistance Test	Airborne Ultrasonic Test Power Factor Test High Voltage Test
Compressors	Vibration Analysis Balance Test and Measurement Alignment (Laser preferred) Lubricating Oil Test Thermodynamic Performance Tests	Hydraulic Oil Test
Cranes	Vibration Analysis Balance Test and Measurement Alignment (Laser preferred) Lubricating Oil Test Mechanical Performance Tests	Insulation Resistance Test Hydraulic Oil Test
Fans	Vibration Analysis Balance Test and Measurement Alignment (Laser preferred) Lubricating Oil Test Thermodynamic Performance Tests	
Gearboxes	Vibration Analysis Hydraulic Oil Test Lubricating Oil Test	
Heat Exchangers	Hydrostatic Test Airborne Ultrasonic Test Thermodynamic Performance Tests	Infrared Thermography

Table A-1
Continued

Equipment Item	Recommended Predictive Technologies	Optional Predictive Technologies
HVAC Ducts	Operational Test Ductwork Leakage Test	
Motor Control	Airborne Ultrasonic Test	Insulation Resistance Test
Centers	Infrared Thermography	
Switchgear	Airborne Ultrasonic Test Insulation Resistance Test Infrared Thermography	Contact Resistance Test High Voltage Test Power Factor Test
Motors	Vibration Analysis Balance Test and Measurement Alignment (Laser preferred) Power Factor Test	Infrared Thermography Insulation Resistance Test Motor Circuit Evaluation Test High Voltage Test
Piping Systems	Hydrostatic Test Thermodynamic Performance Tests	Airborne Ultrasonic Test Pulse Ultrasonic Test Infrared Thermography
Pumps	Vibration Analysis Balance Test and Measurement Alignment (Laser preferred) Lubricating Oil Test Thermodynamic Performance Tests	Hydraulic Oil Test
Roofs, Walls and Insulation	Infrared Thermography	Airborne Ultrasonic Test
Steam Traps	Airborne Ultrasonic Test	
Transformers	Airborne Ultrasonic Test Power Factor Test Insulation Oil Test Infrared Thermography Turns Ratio Test	Contact Resistance Test Insulation Resistance Test High Voltage Test
Valves	Hydrostatic Test	Airborne Ultrasonic Test Thermodynamic Performance Tests Infrared Thermography

Failure Analysis Form

Reliability Centered Maintenance

Equipment Identification: Item No. _____ Equip. Type _____

                           Ser. No. _____ Location _____

Name of Person(s) Responding: _____

Time of Equipment Failure: Date: _____ Time: _____

Time Equipment Returned to Service: Date: _____ Time: _____

Brief Description of Failure: _____

_____

_____

Probable Cause of Failure: _____

_____

_____

Corrective Action Taken: _____

_____

_____

Parts Replaced: _____

_____

Previous Failures (review CMMS): _____

Date Last PM was Performed: _____ Associated W.O. No.: _____

Direct Cost Data: In-House   Contract   Subtotal

           Labor: $_____ $_____ $_____

        Material: $_____ $_____ $_____

                          Total Cost: $_____

Failure Analysis Report Completed By: _____ Date: _____

Figure A-2 Failure Analysis Form

Sample CMMS Data Collection Form

Initials: _____

Film/Roll/Frame: _____   Date: _____

Mechanical Component: Equipment #: _____ Belongs to: _____
Equipment: _____
Location: _____ Floor: _____ Room _____
MFG: _____

Model, ID, Spec, No.: _____ Size: _____
Model, ID, Spec, No.: _____
Model, ID, Spec, No.: _____ Type: _____
Model, ID, Spec, No.: _____

SN # 1: _____ SN # 2: _____
SN # 3: _____ SN # 4: _____

RPM: _____ GPM: _____ CFM/FPM: _____ MWP @ °F: _____

Oil, Air, Fuel Filter Type: _____ MN, PN, Size: _____ Qty.: _____

Oil, Air, Fuel Filter Type: _____ MN, PN, Size: _____ Qty.: _____

Oil, Air, Fuel Filter Type: _____ MN, PN, Size: _____ Qty.: _____

Oil, Air, Fuel Filter Type: _____ MN, PN, Size: _____ Qty.: _____

Oil, Air, Fuel Filter Type: _____ MN, PN, Size: _____ Qty.: _____

Lubricant: Type Oil _____ Type Grease _____

Drive Type: Belt, Chain, Gear, Coupling, Clutch, Direct: _____

Belt Type/Size/Qty. ( ): _____ Coupling Type/Size: _____

Remarks: _____
_____
_____

Electrical Component: Equipment #: _____ Belongs to: _____
Equipment: _____
Location: _____ Floor: _____ Room: _____
MFG: _____

Model, ID, Spec, Cat: _____
Model, ID, Spec, Cat: _____
Model, ID, Spec, Cat: _____
Model, ID, Spec, Cat: _____

SN # 1: _____ SN # 2: _____
SN # 3: _____ SN # 4: _____
HP: _____ Kva/kW: _____ RPM: _____ Amps: _____ V/Ph AC DC: _____ Frame: _____
HP: _____ Kva/kW: _____ RPM: _____ Amps: _____ V/Ph AC DC: _____ Frame: _____
Bearings: Greased: ____ Sealed: ____ Oil Fed: ____ Type Oil: ____ Type Grease: _____
Remarks:

Figure A-3 Sample CMMS Data Collection Form

Enter name of company being certified here

PHASE	CATEGORY	SUB CATEGORY	ITEM NAME	MAXIMUM SCORE	THIS SCORE	LOW	LOW AVERAGE	AVERAGE	HIGH AVERAGE	HIGH
			LEAN MAINTENANCE PREPARATION, IMPLEMENTATION AND EXECUTION AUDIT					SCORING		
1			**PREPARATION AND PLANNING PHASE**	25.00	18.75					
	1.1	-	Management Commitment	4.00	3.00				x	
	1.2	-	Project Manager Assignment	4.00	3.00				x	
	1.3	-	Planning							
		1.3.1	Planning Meeting							
		1.3.1.1	Appropriate Attendees	4.00	3.00				x	
		1.3.1.2	Agenda and Action Items	4.00	3.00				x	
		1.3.2	Master Plan Approval / Published							
		1.3.2.1	Schedule and Milestones	4.00	3.00				x	
		1.3.2.2	Mission Statement	4.00	3.00				x	
		1.3.2.3	Objectives and Goals	4.00	3.00				x	
		1.3.2.4	Assignment of Responsibilities	4.00	3.00				x	
		1.3.2.5	Completeness of Plan	4.00	3.00				x	
	1.4	-	Pilot Project Selected and Planned	4.00	3.00				x	
	1.5	-	Project Selling Campaign / Training Established	4.00	3.00				x	
2	-	-	**IMPLEMENTATION PHASE**	28.57	21.43					
	2.1	-	Organization							
		2.1.1	Structure / Infrastructure Planned	4.00	3.00				x	
		2.1.2	Work Flow Planning	4.00	3.00				x	
	2.2	-	CMMS							
		2.2.1	CMMS Selection	4.00	3.00				x	
		2.2.2	CMMS Implementation							
		2.2.2.1	Equipment Inventory Complete and Accurate	4.00	3.00				x	
		2.2.2.2	Technical / Procedural Documentation in CMMS	4.00	3.00				x	
		2.2.2.3	Reports Generation Complete	4.00	3.00				x	
		2.2.2.4	Work Order Sequence Established and Followed	4.00	3.00				x	
	2.3	-	Total Productive Maintenance							
		2.3.1	PM / PdM / Condition Monitoring & Testing Program	4.00	3.00				x	
		2.3.2	5-S / Visual Responsibility & Planning	4.00	3.00				x	
		2.3.3	Work Order System Complete	4.00	3.00				x	
		2.3.4	Planner / Scheduler							
		2.3.4.1	Qualified Assignee	4.00	3.00				x	
		2.3.4.2	Training	4.00	3.00				x	
		2.3.5	Value Stream Process Planning	4.00	3.00				x	
	2.4	-	Maintenance Engineering (Est. & Assign Responsibilities)							
		2.4.1	Responsibilities Assigned / Understood / Practiced	4.00	3.00				x	
	2.5	-	MRO Storeroom							
		2.5.1	Lean Policies and Controls in Place	4.00	3.00				x	
		2.5.2	Reorganization completed	4.00	3.00				x	
		2.5.3	Supplier	4.00	3.00				x	
	2.6	-	Training							
		2.6.1	General Lean Maintenance Training Completed	4.00	3.00				x	
		2.6.2	JTA and SA in progress or completed	4.00	3.00				x	
3	-	-	**EXECUTION PHASE**	46.43	34.82					
	3.1	-	Organization							
		3.1.1	Integration	4.00	3.00				x	
		3.1.2	Communications	4.00	3.00				x	
		3.1.3	Work Flow Discipline	4.00	3.00				x	
	3.2	-	CMMS							
		3.2.1	Complete and Effective Use	4.00	3.00				x	
		3.2.2	Reporting Effectiveness	4.00	3.00				x	
		3.2.3	Work Order Discipline	4.00	3.00				x	
	3.3	-	Total Productive Maintenance (TPM)							
		3.3.1	Use of Documentation	4.00	3.00				x	
		3.3.2	CMMS Implementation							
		3.3.2.1	Equipment Inventory	4.00	3.00				x	
		3.3.2.2	Technical / Procedural Documentation	4.00	3.00				x	
		3.3.2.3	Report Generation	4.00	3.00				x	
		3.3.2.4	Effective Scheduler Usage	4.00	3.00				x	
		3.3.3	5-S / Visual Deployment	4.00	3.00				x	
		3.3.4	Work Measurement	4.00	3.00				x	
		3.3.5	Work Order Usage	4.00	3.00				x	
		3.3.6	Schedule Compliance	4.00	3.00				x	
		3.3.7	Value Stream Mapping Process	4.00	3.00				x	
	3.4	-	Maintenance Engineering							
		3.4.1	Knowledge of Lean Maintenance Tools	4.00	3.00				x	
		3.4.2	Predictive Maintenance / Condition Monitor & Testing	4.00	3.00				x	
		3.4.3	Continuous PM Evaluation & Improvement	4.00	3.00				x	
		3.4.4	Planning & Scheduling Process	4.00	3.00				x	
	3.5	-	MRO Storeroom							
		3.5.1	Inventory Control	4.00	3.00				x	
		3.5.2	Storeroom Organization	4.00	3.00				x	
		3.5.3	CMMS Integration / Work Order Use	4.00	3.00				x	
		3.5.4	Stockouts	4.00	3.00				x	
		3.5.5	Use of JIT Vendors	4.00	3.00				x	
	3.6	-	Training							
		3.6.1	Skill needs / Skills availability assessed	4.00	3.00				x	
		3.6.2	Focused Training Program	4.00	3.00				x	
		3.6.3	Qualification / Certification Program	4.00	3.00				x	
		3.6.4	Training Effectiveness Evaluation	4.00	3.00				x	
	3.7	-	Lean Sustainment Established	4.00	3.00				x	
			TOTALS	100.00	75.00					

Figure A-4 Lean Maintenance Preparation, Implementation and Execution Audit

CATEGORY	SUB CATEGORY	ITEM NAME	MAXIMUM SCORE	THIS SCORE	LOW	LOW AVERAGE	AVERAGE	HIGH AVERAGE	HIGH
1	-	PROGRAM FOUNDATION	21.62	10.14					
	1a	Management Commitment	4.00	3.00				x	
	1b	Master Plan	4.00	2.00			x		
	1c	Mission Statement	4.00	0.00	x				
	1d	Objectives and Goals	4.00	2.00			x		
	1e	Management Reporting	4.00	1.00		x			
2	-	ORGANIZATION	10.81	5.41					
	2a	Structure	4.00	1.00		x			
	2b	Work Flow	4.00	2.00			x		
	2c	Communication	4.00	3.00				x	
3	-	TOTAL PRODUCTIVE MAINTENANCE	29.73	17.57					
	3a	Standardized Work Practices	4.00	2.00			x		
	3b	CMMS Implementation							
	3b(1)	Equipment Inventory	4.00	3.00				x	
	3b(2)	Technical / Procedural Documentation	4.00	3.00				x	
	3b(3)	Report Generation	4.00	3.00				x	
	3c	5-S / Visual Deployment	4.00	2.00			x		
	3d	Work Measurement	4.00	2.00			x		
	3e	Work Order Usage	4.00	3.00				x	
	3f	Schedule Compliance	4.00	2.00			x		
	3g	Value Stream	4.00	2.00			x		
4	-	MAINTENANCE ENGINEERING	13.51	9.46					
	4a	Knowledge of Lean Maintenance Tools	4.00	2.00			x		
	4b	Predictive Maintenance / Condition Monitor & Testing	4.00	3.00				x	
	4c	Continuous PM Evaluation & Improvement	4.00	3.00				x	
	4d	Planning & Scheduling	4.00	3.00				x	
5	-	MRO STOREROOM	13.51	4.73					
	5a	Inventory Control	4.00	3.00				x	
	5b	Storeroom Organization	4.00	2.00			x		
	5c	CMMS Integration / Work Order Use	4.00	2.00			x		
	5d	Stockouts	4.00	1.00		x			
	5e	Use of JIT Vendors	4.00	2.00			x		
6	-	TRAINING	10.81	8.11					
	6a	Skill needs / Skills availability assessed	4.00	2.00			x		
	6b	Focused Training Program	4.00	2.00			x		
	6c	Qualification / Certification Program	4.00	4.00					x
	6d	Training Effectiveness Evaluation	4.00	4.00					x
		TOTALS	100.00	55.41					

Figure A-5 Lean Maintenance Practices Audit

Table A-2
Predictive Maintenance Data Collection Forms—1

Vibration Analysis Test Criteria

Item	Date of Inspection	Acceptable Limit	Actual Value	Inspector Initials	PASS	FAIL	Comments
Vibration Analysis Test							
Test Instrumentation							
FFT Analyzer							
Type							
Model							
Serial Number							
Last Calibration Date							
Line Resolution Bandwidth							
Dynamic Range							
Hanning Window							
Linear Non-overlap Averaging							
Anti-aliasing Filters							
Amplitude Accuracy							

Table A-2
Continued

Sound Disk Thickness						
Adhesive (hard/soft)						
Vibration Readings						
1H						
1V						
1A						
2H						
2V						
2A						
Velocity Amplitude (in./sec.—peak)						
Running Speed Order						
Acceleration Overall Amp (g—peak)						
Vibration Signatures (H, V, A)						
Frequency (CPM)						
Balanced Condition						
Balance Wt. Type						
Results:						

Table A-3
Predictive Maintenance Data Collection Forms—2

Lubrication Oil Test Criteria

Item	Date of Inspection	Acceptable Limit	Actual Value	Inspector Initials	PASS	FAIL	Comments
Lubrication Oil Test							
Liquids							
Viscosity Grade (ISO Units)							
AGMA/SAE Classification							
Additives							
Grease							
Type of Base Stock							
NLGI Number							
Type/% of thickener							
Dropping Point							
Base Oil Viscosity (SUS)							
Total Acid Number							
Visual Observation (Cloudiness)							
IR Spectral Analysis—Metal Count							
Particle Count							
Water Content							
Viscosity							
Results:							

Table A-4

Predictive Maintenance Data Collection Forms—3

Alignment Criteria

Item	Date of Inspection	Acceptable Limit	Actual Value	Inspector Initials	PASS	FAIL	Comments
Alignment							
RPM							
Soft Foot Actual (in.)							
Soft Foot Tolerance							
Vert. angularity at coupling—Actual							
Vert. offset at coupling—Actual							
Vert. angularity at coupling—Actual							
Vert. offset at coupling—Actual							
Axial Shaft Play							
Shims							
Shim Type							
Shim Condition							
Number of Shims in Pack							
Thickness							
Sheaves							
True to Shaft							
Runout (in.)							
Results:							

Table A-5
Predictive Maintenance Data Collection Forms—4

Breaker Timing Test Criteria

Item	Date of Inspection	Acceptable Limit	Actual Value	Inspector Initials	PASS	FAIL	Comments
Breaker Timing Test							
Voltage Applied							
C1—Phase A							
C2—Phase B							
C3—Phase C							
Results:							

Table A-6
Predictive Maintenance Data Collection Forms—5

Contact Resistance Test Criteria

Item	Date of Inspection	Acceptable Limit	Actual Value	Inspector Initials	PASS	FAIL	Comments
Contact Resistance Test							
DC Current Applied							
Measured Voltage							
Calculated Resistance							
Manufacturer Resistance							
Results:							

Table A-7
Predictive Maintenance Data Collection Forms—6

Insulation Oil Test Criteria

Item	Date of Inspection	Acceptable Limit	Actual Value	Inspector Initials	PASS	FAIL	Comments
Insulation Oil Test							
Dissolved Gas Analysis							
Nitrogen (N2)		<100 ppm					
Oxygen (O2)		<10 ppm					
Carbon Dioxide (CO2)		<10 ppm					
Carbon Monoxide (CO)		<100 ppm					
Methane (CH4)		None					
Ethane (C2H6)		None					
Ethylene (C2H4)		None					
Hydrogen (H2)		None					
Acetylene (C2H2)		None					
Karl Fisher (@ 20 Deg. C)		<25 ppm					

Table A-7
Continued

Dielectric Breakdown	>30kV				
Strength					
Neutralization Number	<0.05 mg/g				
Interfacial Tension	>40 dynes/cm				
Color (ASTM D-1524)	<3.0				
Sediment	clear				
Power Factor	<0.05%				
Sediment/Visual Examination	clear				
Results:					

Table A-8
Predictive Maintenance Data Collection Forms—7

Power Factor Test Criteria

Item	Date of Inspection	Acceptable Limit	Actual Value	Inspector Initials	PASS	FAIL	Comments
Power Factor Test							
Grounded Specimen Test—GST							
Ungrounded Specimen Test—UST							
GST with Guard							
Environment Humidity							
Environment Temperature							
Surface Cleanliness							
Phase I							
Applied Voltage							
Total Current							
Capacitive Current							

Continued

Dissipation Factor								
Power Factor								
Normal Power Factor								
Phase II								
Applied Voltage								
Total Current								
Capacitive Current								
Dissipation Factor								
Power Factor								
Normal Power Factor								
Phase III								
Applied Voltage								
Total Current								
Capacitive Current								
Dissipation Factor								
Power Factor								
Normal Power Factor								
Results:								

Procedure for performing a Failure Modes and Effects Analysis (FMEA)

FMEAs are generally performed using the guidance provided in MIL-STD-1629A, in spite of the fact that, in theory, Military Standards no longer exist. A functional process based on MIL-STD-1629A is outline here.

1. Describe, in words, the process and functions of the system/equipment to be analyzed. This step is really meant to ensure that the analyst has a clear understanding of what the equipment is meant to do and how it fits into the overall production scheme. There is no need to create an eloquent thesis. Just write down short "one-liners" that describe the various functions and the overall system process.
2. As you refine and put order to the written descriptions, begin creating a diagram of the process which basically will consist of ordered blocks representing the various functions. If not previously defined, system boundaries will need to be established. It may help in some cases to sketch a pictorial representation in order to better visualize components and their functions. When completed, the block diagram will need to be completed "smooth" as it will be a permanent attachment to the FMEA Data Form as in Figure A-6.
3. On a rough copy of the FMEA data form, begin listing the functions as identified above.
4. Identify functional failures or failure modes. Note that a failure mode in one component can be a cause of a failure mode in another component. In some cases the iterations of failure modes and causes can be extensive. All failure modes should be identified, regardless of their probability of occurring (see Figure A-7).
5. Describe the effects of each failure mode and assign a severity ranking. Effects can be on the component, on the next step in the process (or block in the system diagram), on the end result of the process or all three. Be sure to consider safety and environmental effects as well as effects on production and product. If not previously established you will need to develop a ranking system for the severity of the effect. This is normally a 1 to 10 scale where 10 is the most severe and 1 indicates none or negligible severity.
6. Identify potential causes of each failure mode. Be sure to consider all possibilities, including poor design, extremes of operating environments, operator (or maintenance tech) error—which in turn may be due to inadequate training, documentation or procedure errors, etc.

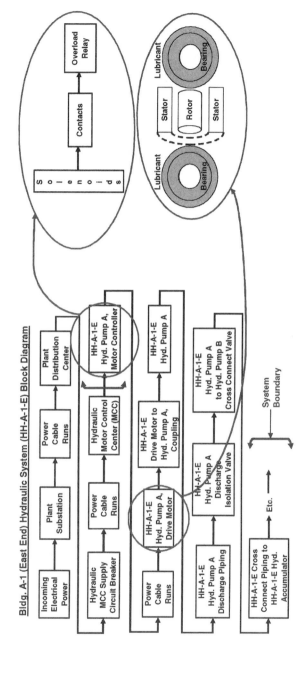

Figure A-6 Block diagram

Hydraulic System HH-A-1-E

Function	Functional Failure	Failure Mode	Source of Failure
Provide hydraulic oil at specified Pressure and Demand volume (flow rate)	Total loss of pressure, volume and flow	Electric Motor Failed Pump Failed Major Leak Blocked Line Valve out of position	(See next table)

Electric Motor

Function	Functional Failure	Failure Mode	Source of Failure
Drive hydraulic pump at specified power level	Motor will not turn	Insulation failure in stator Insulation failure in rotor Bearing Seized Motor Controller Contactor Failed Circuit Breaker tripped	(See next table)

Motor Bearings

Failure Mode	Mechanism	Reason	Cause
Bearing seized (this includes seals, shields, lubrication system, and lock nuts)	Lubrication	Contamination	Seal failed
			Supply dirty
		Wrong Type	Procedure wrong
			Supply information wrong
		Too Little	Leak
			Human error
			Procedure error
		Too Much	Human error
			Procedure error
	Fatigue	Metallurgical	Inherent
			Excessive temperature
		Excessive Load	Mechanical imbalance
			Misalignment
			Wrong application (bearing not sized for the load)
	Etc.		

Figure A-7 Hydraulic System

7. Enter the probability factor for each potential cause. (refer to Table 3-7 in Section 3.2.2.2)
8. Identify the compensating provisions, which are either design or process controls intended to (a) prevent the cause of a failure from

Detection	Rank
Almost Impossible	10
Very Remote	9
Remote	8
Very Low	7
Low	6
Moderate	5
Moderately High	4
High	3
Very High	2
Almost Certain	1

Figure A-8 Detection

occurring or (b) identify the potential for the cause to occur (i.e., existing PdM procedure).

9. Determine the likelihood of detection. Detection is an assessment of the likelihood that the compensating provisions (design or process) will detect the Cause of the Failure Mode or the Failure Mode itself. If the Compensating Provisions include an existing PdM procedure, the Likelihood of Detection is the likelihood that the potential for the Cause to occur will be detected. (see Figure A-8).

10. Calculate and enter the Failure Mode Ranking. The ranking is the mathematical product of the numerical Severity, Probability and Detection ratings. Ranking = (Severity) × (Probability) × (Detection). The ranking is used to prioritize those items requiring additional action.

11. Enter any remarks pertinent to the FMEA item.

Management Oversight and Risk Tree (MORT) Analysis

A Mini-MORT analysis chart is shown at the end of this discussion. This chart is a checklist of what happened (less-than-adequate specific barriers and controls) and why it happened (less-than-adequate management).

To perform the MORT analysis:

1. Identify the problem associated with the occurrence and list it as the top event.

Failure Mode and Effects Analyis
(System Name)

System: _____
Indenture Level: _____
Reference Drawing: _____
Mission: _____

Date: _____
Page ___ of ___
Compiled by: _____
Approved by: _____

Identification Number	Item Functional Identification (Nomenclature)	Function	Failure Modes and Causes	Probability Factor	Failure Effects			Likelihood of Failure Detection	Compensating Provisions	Severity Class	Ranking	Remarks
					Local Effects	Next Higher Level	End Effects					

Figure A-9 Failure Mode and Effects Analysis

METHOD	WHEN TO USE	ADVANTAGES	DISADVANTAGES	REMARKS
Events and Causal Factor Analysis	Use for multi-faceted problems with long or complex causal factor chain.	Provides visual display of analysis process. Identifies probable contributors to the condition.	Time-consuming and requires familiarity with process to be effective.	Requires a broad perspective of the event to identify unrelated problems. Helps to identify where deviations occurred from acceptable methods.
Change Analysis	Use when cause is obscure. Especially useful in evaluating equipment failures.	Simple six-step process.	Limited value because of the danger of accepting wrong, "obvious" answer.	A singular problem technique that can be used in support of a larger investigation. All root causes may not be identified.
Barrier Analysis	Use to identify barrier and equipment failures and procedural or administrative problems.	Provides systematic approach.	Requires familiarity with process to be effective.	This process is based on the MORT Hazard/Target Concept.
MORT/Mini-MORT	Use when there is a shortage of experts to ask the right questions and whenever the problem is a recurring one. Helpful in solving programmatic problems.	Can be used with limited prior training. Provides a list of questions for specific control and management factors.	May only identify area of cause, not specific causes.	If this process fails to identify problem areas, seek additional help or use cause-and-effect analysis.
Human Performance Evaluations (HPE)	Use whenever people have been identified as being involved in the problem cause.	Thorough analysis.	None if process is closely followed.	Requires HPE training.
Kepner-Tregoe	Use for major concerns where all aspects need thorough analysis.	Highly structured approach focuses on all aspects of the occurrence and problem resolution.	More comprehensive than may be needed.	Requires Kepner-Tregoe training.

Figure A-10 Summary of Root Cause Failure Analysis Methods

2. Identify the elements on the "what" side of the tree that describe what happened in the occurrence (what barrier or control problems existed).
3. For each barrier or control problem, identify the management elements on the "why" side of the tree that permitted the barrier control problem.
4. Describe each of the identified inadequate elements (problems) and summarize your findings.

A brief explanation of the "what" and "why" may assist in using mini-MORT for causal analyses.

When a target inadvertently comes in contact with a hazard and sustains damage, the event is an accident. A hazard is any condition, situation or activity representing a potential for adversely affecting economic values or the health or quality of people's lives. A target can be any process, hardware, people, the environment, product quality or schedule—anything that has economic or personal value.

What prevents accidents or adverse programmatic impact events?

- Barriers that surround the hazard and/or the target and prevent contact or controls and procedures that ensure separation of the hazard from the target.
- Plans and procedures that avoid conflicting conditions and prevent programmatic impacts. In a facility, what functions implement and maintain these barriers, controls, plans and procedures?
- Identifying the hazards, targets, and potential contacts or interactions and specifying the barriers/controls that minimize the likelihood and consequences of these contacts.
- Identifying potential conflicts/problems in areas such as operations, scheduling or quality and specifying management policy, plans and programs that minimize the likelihood and consequences of these adverse occurrences.
- Providing the physical barriers: designing, installation, signs/warnings, training or procedures.
- Providing planning/scheduling, administrative controls, resources or constraints.
- Verifying that the barriers/controls have been implemented and are being maintained by operational readiness, inspections, audits, maintenance and configuration/change control.
- Verifying that planning, scheduling and administrative controls have been implemented and are adequate.

- Policy and policy implementation (identification of requirements, assignment of responsibility, allocation of responsibility, accountability, vigor and example in leadership and planning).

Definitions used with this method:

- A *cause* (causal factor) is any weakness or deficiency in the barrier/ control functions or in the administration/management functions that implement and maintain the barriers/controls and the plans/ procedures.
- A *causal factor chain* (sequence or series) is a logical hierarchal chain of causal factors that extends from policy and policy implementation through the verification and implementation functions to the actual problem with the barrier/control or administrative functions.
- A *direct cause* is a barrier/control problem that immediately preceded the occurrence and permitted the condition to exist or adverse event to occur. Since any element on the chart can be an occurrence, the next upstream condition or event on the chart is the direct cause and can be a management factor. (Management is seldom a direct cause for a real-time loss event such as injury or property damage but may very well be a direct cause for conditions.)
- A *root cause* is the fundamental cause, which, if corrected, will prevent recurrence of this and similar events. This is usually not a barrier/control problem but a weakness or deficiency in the identification, provision or maintenance of the barriers/controls or the administrative functions. A root cause is ordinarily control-related involving such upstream elements as management and administration. In any case, it is the original or source cause.
- A *contributing cause* is any cause that had some bearing on the occurrence, on the direct cause, or on the root cause but is not the direct or the root cause.

Answer the following (and any related items unique to your particular operation) questions in order to fill out the Change Analysis Worksheet:

WHAT?
What is the condition?
What occurred to create the condition?
What occurred prior to the condition?
What occurred following the condition?

What activity was in progress when the condition occurred?
What activity was in progress when the condition was identified?
Operational evolution in the work space?
Surveillance test?
Starting/stopping equipment?
Operational evolution outside the work space?
Valve line-up?
Removing equipment from service?
Returning equipment to service?
Maintenance activity?
Surveillance?
Corrective maintenance?
Modification installation?
Troubleshooting?
Training activity?
What equipment was involved in the condition?
What equipment initiated the condition?
What equipment was affected by the condition?
What equipment mitigated the condition?
What is the equipment's function?
How does it work?
How is it operated?
What failed first?
Did anything else fail due to the first problem?
What form of energy caused the equipment problem?
What are recurring activities associated with the equipment?
What corrective maintenance has been performed on the equipment?
What modifications have been made to the equipment?
What system or controls (barriers) should have prevented the condition?
What barrier(s) mitigated the consequences of the condition?
WHEN?
When did the condition occur?
What was the facility's status at the time of occurrence?
When was the condition identified?
What was the facility's status at the time of identification?
What effects did the time of day have on the condition? Did it affect:
Information availability?
Personnel availability?
Ambient lighting?
Ambient temperature?

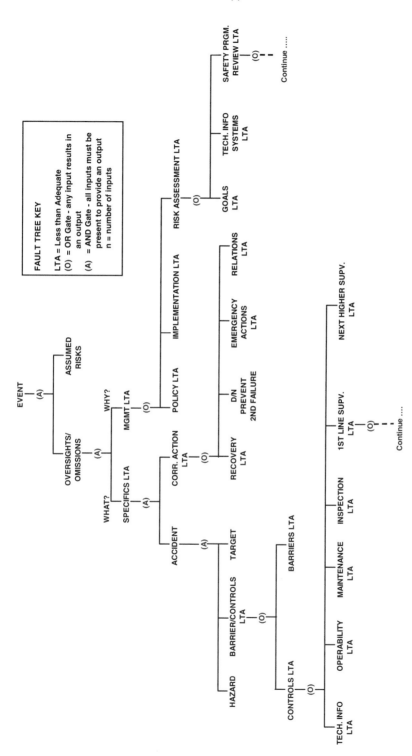

Figure A-12 Management Oversight and Risk Tree (example)

MORT BASED ROOT CAUSE ANALYSIS FORM

		Policy	Policy Implementation					Risk Assessment					Bridge Elements					Specific Factors					Task Performance					
		1	2	3	4	5	6	7	8	9	10	11	12	13	14	15	16	17	18	19	20	21	22	23	24	25	26	27	28	29	30	31
		Policy	Line Responsibility	Staff Responsibility	Accountability	Vigor/Example	Methods/Criteria Analysis	Open	Goals	Technical Info Systems	Hazard Analysis Process	Program Review	Open	Management Services	Directives	Budget	Information Flow	Open	Open	Technical Information	Operational Readiness	Maintenance	Inspection	Supervision	Higher Supervisory Services	Open	Task Assignment	Task Safety Analysis	Procedural Incompatibility	Personnel Performance Discrep.	Open	Open

Findings or Conclusions

Change Factor	Difference/Change	Effect	Questions to Answer
What (conditions, occurrence, activity, equipment)			
When (occurred, identified, plant status, schedule)			
Where (physical location, environmental conditions)			
How (work practice, omission, extraneous action, out of sequence procedure)			
Who (personnel involved, training, qualification, supervision)			

Company _____   Date _____
Plant _____   Name of Person Performing Analysis
                                     _____

Occurrence Description _____

Figure A-14 Change Analysis Worksheet

Did the condition involve shift-work personnel? If so:
  What type of shift rotation was in use?
  Where in the rotation were the personnel?
For how many continuous hours had any involved personnel been working?
WHERE?
Where did the condition occur?
What were the physical conditions in the area?
Where was the condition identified?
Was location a factor in causing the condition?
  Human factor?
    Lighting?
    Noise?
    Temperature?
    Equipment labeling?
    Radiation levels?
    Personal protective equipment required in the area?
    Radiological protective equipment required in the area?

Accessibility?
Indication availability?
Other activities in the area?
What position is required to perform tasks in the area?
Equipment factor?
Humidity?
Temperature?
Cleanliness?
HOW?
Was the condition an inappropriate action or was it caused by an inappropriate action?
An omitted action?
An extraneous action?
An action performed out of sequence?
An action performed to too small of a degree? To too large of a degree?
Was there an applicable procedure?
Was the correct procedure used?
Was the procedure followed?
Followed in sequence?
Followed "blindly"—without thought?
Was the procedure:
Legible?
Misleading?
Confusing?
An approved, current revision?
Adequate to do the task?
In compliance with other applicable codes and regulations?
Did the procedure:
Have sufficient detail?
Have sufficient warnings and precautions?
Adequately identify techniques and components?
Have steps in the proper sequence?
Cover all involved systems?
Require adequate work review?
WHO?
Which personnel:
Were involved with the condition?
Observed the condition?
Identified the condition?
Reported the condition?

Corrected the condition?
Mitigated the condition?
Missed the condition?
What were:
The qualifications of these personnel?
The experience levels of these personnel?
The work groups of these personnel?
The attitudes of these personnel?
Their activities at the time of involvement with the condition?
Did the personnel involved:
Have adequate instruction?
Have adequate supervision?
Have adequate training?
Have adequate knowledge?
Communicate effectively?
Perform correct actions?
Worsen the condition?
Mitigate the condition?

Barrier Analysis Description

There are many things that should be addressed during the performance of a Barrier Analysis. NOTE: In this usage, a barrier is from Management Oversight and Risk Tree (MORT) terminology and is something that separates an affected component from an undesirable condition/situation. The figure at the end of this description provides an example of Barrier Analysis. The questions listed below are designed to aid in determining what barrier failed, thus resulting in the occurrence.

What barriers existed between the second, third, etc. condition/
situation and the second, third, etc. problems?
If there were barriers, did they perform their functions? Why?
Did the presence of any barriers mitigate or increase the occurrence
severity? Why?
Were any barriers not functioning as designed? Why?
Was the barrier design adequate? Why?
Were there any barriers in the condition/situation source(s)? Did they
fail? Why?
Were there any barriers on the affected component(s)? Did they fail?
Why?
Were the barriers adequately maintained?

Were the barriers inspected prior to expected use?

Why were any unwanted energies present?

Is the affected system/component designed to withstand the condition/situation without the barriers? Why?

What design changes could have prevented the unwanted flow of energy? Why?

What operating changes could have prevented the unwanted flow of energy? Why?

What maintenance changes could have prevented the unwanted flow of energy? Why?

Could the unwanted energy have been deflected or evaded? Why?

What other controls are the barriers subject to? Why?

Was this event foreseen by the designers, operators, maintainers, anyone?

Is it possible to have foreseen the occurrence? Why?

Is it practical to have taken further steps to have reduced the risk of the occurrence?

Can this reasoning be extended to other similar systems/components?

Were adequate human factors considered in the design of the equipment?

What additional human factors could be added? Should be added?

Is the system/component user friendly?

Is the system/component adequately labeled for ease of operation?

Is there sufficient technical information for operating the component properly? How do you know?

Is there sufficient technical information for maintaining the component properly? How do you know?

Did the environment mitigate or increase the severity of the occurrence? Why?

What changes were made to the system/component immediately after the occurrence?

What changes are planned to be made? What might be made?

Have these changes been properly and adequately analyzed for effect?

What related changes to operations and maintenance have to be made now?

Are expected changes cost effective? Why? How do you know?

What would you have done differently to have prevented the occurrence, disregarding all economic considerations (as regards operation, maintenance and design)?

What would you have done differently to have prevented the occurrence, considering all economic concerns (as regards operation, maintenance and design)?

Work Task: Clean Control Relay Panel and Contacts

Occurrence: Production Line #2 Power Trip

<u>Sequence of Events:</u>

System Tagout Requested → Warning Tag Hung → Maintenance Electricians Given Assignment → Electricians Follow Procedure → Line #2 Power Trip

<u>Barrier Analysis:</u>

Start of Work Process → Tagout Process Step 1 → Tagout Process Step 2 → Communications Process Interface → Procedure

Occurrence ← Training

Maint. requests de-energizing two panels so relays can be cleaned. Operations will only allow one panel at a time to be tagged out. Electrical supervisor told and agrees.	Tag hung on P689 - only P690 is still energized.	Electricians given W.O. to work, which references a Maint. Procedure, but not told of change in scope by supervisor.	Electricians go to P690 and begin procedure. Procedure has no step to verify dead power supply before starting. They open first relay and Line #2 trips.	Electricians never trained to always check power supply prior to working on electrical equipment.
Barrier Holds	**Barrier Holds**	**Barrier Fails**	**Barrier Fails**	**Barrier Fails**
Barrier Holds	**Barrier Holds**	**Barrier Fails**	**Barrier Fails**	**Barrier Fails**

Figure A-15 Work Task

Approximating Failure Distributions

The four failure rate functions or hazard functions corresponding to the probability density functions (exponential, Weibull, lognormal and normal) are shown in Figure A-16.

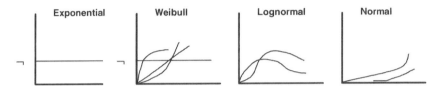

Figure A-16 Failure Rate Functions

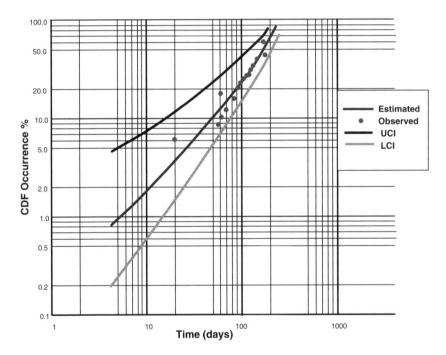

Figure A-17 Weibull distribution

Appendix B

Documentation Examples

Motor Control Center	OJT No.	330J02
	Revision	Orig.

330.02

01 . . . .

02 . . . .

03 . . . .

04 . . . .

05   Test individual MCC components

_____
Initial / Date

Standard
The electrical characteristics of each of the following components are checked with appropriate testing equipment and determined to be within Vendor and AP specifications

Fuses	Capacitors
Starters	System status indicator lights
Transformers	

Test results documented on Work Order or other applicable documentation
Components determined to be out of specification are indicated on the Work Order

Training / Reference Material
AP Standard Practices, Vendor/OEM manuals, AP Engineering Specifications, Electrical Schematics
TPC Unit 210, Lesson One, _Troubleshooting with Electrical Schematics_
TPC Unit 210, Lesson Four, _Troubleshooting Combination Starters_
TPC Unit 210, Lesson Five, _Troubleshooting Control Devices_

06 . . . .

Figure B-1 Sample of an OJT Training Guide

TSA 201 - Centrifugal Pumps	Exam No.	201E01
EXAMINATION	Revision No.	Orig.

1. Centrifugal Pumps are used to:
   A. mix fluids with different specific gravities
   B. move fluids through a piping system
   C. compress gases
   D. compress liquids

   .
   .
   .

23. The alignment shown here is known as:
    A. vertical angular alignment
    B. horizontal angular alignment
    C. vertical parallel alignment
    D. horizontal parallel alignment

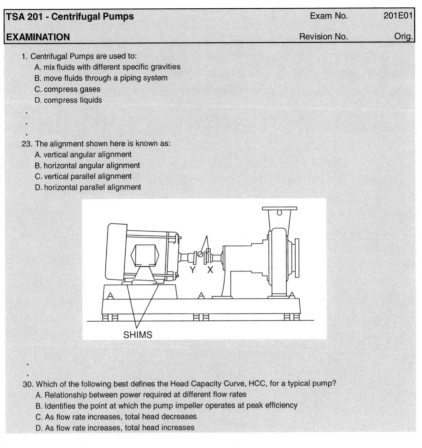

   .
   .

30. Which of the following best defines the Head Capacity Curve, HCC, for a typical pump?
    A. Relationship between power required at different flow rates
    B. Identifies the point at which the pump impeller operates at peak efficiency
    C. As flow rate increases, total head decreases
    D. As flow rate increases, total head increases

Figure B-2 Questions From a Sample Examination

TQS COMPLETION & VERIFICATION RECORD	Document No.	101CV00
	Revision	Orig.
101 - GENERAL KNOWLEDGE AND SKILLS	Date	01/01/02
	Page	1 of 1

EMPLOYEE NAME: _____   EMPLOYEE NO. _____

INSTRUCTIONS

Each of the following items must be completed in preparation for the written examination. The trainee shall initial and date
each element as completed. The trainee understands that each item listed under TSA headings are completed when initials
are provided.

All knowledge items must be completed before the written examination is administered. The written examination must be
passed before formal on-the-job training commences. The Maintenance Manager shall initial and date adjacent to each
examination requirement indicating that the trainee has successfully completed and passed the required examination.

Classroom Instruction and On-the-job Training Guides are referenced and initialed by the instructor/coach after completion and/or
sufficient practice has been provided and the trainee has demonstrated an acceptable level of proficieny. Once the training
has been initialed, the cognizant Maintenance Supervisor or Coach conducts the Skill Evaluation.

TSA 101 HEADINGS

101.1	Identify common work tools and their function.	t	_____	rainee initial
101.2	Identify common mechanical processes and their purpose.		_____	and date
101.3	.		_____	
101.4	.		_____	
101.5	.		_____	
101.6	.		_____	
101.7	.		_____	
101.8	.		_____	
101.9	.		_____	
101.10	.		_____	
101.11	.		_____	
101.12	Identify key aspects of the facility safety program and worker safety practices.		_____	

MAINTENANCE MANAGER
Written Exam - 101G01   _____ Maint. Mgr.
                                                                          init. & date

When all of the indicated Maintenance Supervisors or Coaches and the Maintenance Manager have initialed this document,
the trainee has completed all required training and examinations and is deemed qualified without restriction for the TSA indicated.

EXPIRATION DATE: _____None_____ (60 month maximum from date of qualification)

Figure B-3 Typical TQS Completion and Verification Record

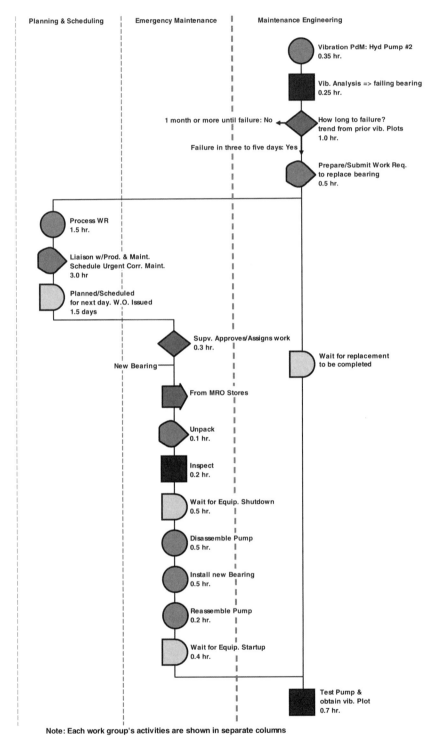

Note: Each work group's activities are shown in separate columns

Figure B-4 Process Mapping Example

EQUIPMENT MAINTENANCE PLAN

EQUIPMENT NUMBER 321300					DESCRIPTION #1 OUTFEED SCREEN CONVEYOR	LOCATION SCREEN/DRYER AREA		DOCUMENTATION AVAILABL YES LIBRARY AND CLEAN ROOM			NAME PLATE VALIDATED YES 6/20/2002
ITEM #	MAINTENANCE TASK DESCRIPTION	FREQUENCY	CRAFT/LEVEL	# CRAFTSMEN	EQUIPMENT CONDITION	TYPE	PROCEDURE / TASK #	EST.TIME (hrs)	SPECIAL TOOLS, MATERIALS & TEST EQUIP.	RELATED MAINTENANCE PROCEDURE(S)	ANNUAL HOURS
001	PERFORM CONDITION MONITORING INSPECTION	W	OP	1	RUNNING	C	CFS CNVYR-W01	0.5	NONE		26
002	CLEAN AND INSPECT TAIL ROLLER AREA	W	OP	1	SHUT DOWN	PM	CFS CNVYR-W02	0.5	NONE		26
003	PERFORM MECHANICAL INSPECTION	M	MECH II	1	SHUT DOWN	PM	CFS CNVYR-M01	1.0	NONE	CFS CNVYR-W02	12
004	PERFORM SAFETY GUARD INSPECTION	M	OP	1	RUNNING	PM	CFS CNVYR-M02	0.5	NONE		6
005	OBTAIN VIBRATION SIGNATURE OF CONVEYOR TAIL ROLLER BEARING	M	MECH I / M.ENG	1 / 1	RUNNING	PdM	CFS CNVYR-M-V01	0.5	ACCELEROMETERS-A1200; RECORDER-R1100; AMP-V112		6
006	REPLACE AUTO LUBRICATORS	Q	MECH II	1	SHUT DOWN	PM	CFS CNVYR-Q01	1.0	P.N. AL-3387 (6 EA)		4
007	PERFORM CONVEYOR BELT AND LACING INSPECTION	Q	MECH II	1	SHUT DOWN	PM	CFS CNVYR-Q02	1.0	NONE		4
008	CHECK DRIVE BELT /CHAIN ALIGNMENT AND TENSION	Q	MECH II	1	SHUT DOWN	PM	CFS CNVYR-Q03	1.0	NONE		4
009	REMOVE AND REPLACE CONVEYOR MAIN DRIVE MOTOR: PERFORM ALIGNMENT	SIT	MECH II / ELEC II	1 / 1	SHUT DOWN	CM	CFS CNVYR-SIT 01	2.0	P.N.SIEMENS 60-450-A		0
010	REMOVE AND REPLACE CONVEYOR MAIN DRIVE GEARBOX END	SIT	MECH II	2	SHUT DOWN	CM	CFS CNVYR-SIT 02	5.0	SEE: CFS CNVYR-SIT 02		0
011	REMOVE AND REPLACE CONVEYOR MAIN DRIVE BELTS	SIT	MECH II	1	SHUT DOWN	CM	CFS CNVYR-SIT 03	1.0	SEE: CFS CNVYR-SIT 03		0
012	REMOVE AND REPLACE CONVEYOR MAIN DRIVE CHAIN & SPROCKETS	SIT	MECH II / MECH I	1 / 1	SHUT DOWN	CM	CFS CNVYR-SIT 04	3.0	SEE: CFS CNVYR-SIT 04		0
013	REMOVE AND REPLACE CONVEYOR BELT	SIT	MECH II	4	SHUT DOWN	CM	CFS CNVYR-SIT 05	6.0	SEE: CFS CNVYR-SIT 05		0
014	REMOVE AND REPLACE HEAD ROLLER BEARING	SIT	MECH II	2	SHUT DOWN	CM	CFS CNVYR-SIT 06	4.0	SEE: CFS CNVYR-SIT 06		0
015	REMOVE AND REPLACE AIR KNIFE BLOWER MOTOR	SIT	MECH II / ELEC II	2 / 1	SHUT DOWN	CM	CFS CNVYR-SIT 07	1.0	SEE: CFS CNVYR-SIT 07		0
016	REMOVE AND REPLACE AIR KNIFE BLOWER FAN	SIT	MECH II	1	SHUT DOWN	CM	CFS CNVYR-SIT 08	1.0	SEE: CFS CNVYR-SIT 08		0
017	REMOVE AND REPLACE CONVEYOR TAIL ROLLER BEARINGS	SIT	MECH II	1	SHUT DOWN	CM	CFS CNVYR-SIT 09	2.0	SEE: CFS CNVYR-SIT 09		0

ANNUAL MAINTENANCE HRS.	88.6
ANNUAL SHUTDOWN HRS.	50.4
ANNUAL OPERATOR HRS.	58.2
ANNUAL MECH HRS.	30.4
ANNUAL ELEC HRS.	0
ANNUAL CONTRACTOR HRS.	0

KEY   FREQUENCY:
W = WEEKLY
M = MONTHLY
Q = QUARTERLY; 2Q = SEMIANNUALLY
A = ANNUALLY
SIT = SITUATIONAL REQUIREMENT

TYPE:
C = CONDITION MONITORING OR TESTING
PM = PREVENTIVE MAINTENANCE
CM = CORRECTIVE MAINTENANCE
PdM = PREDICTIVE MAINTENANCE

Figure B-5 Equipment Maintenance Plan (EMP) Example

CRITICALITY ANALYSIS of HVAC SYSTEM

A Asset	B Equipment	C Description	D Location	E Function Location	F Acquisition Value	G Mission Impact	H Safety Impact	I Environmental Impact	J Single Point Failure	K Preventive Maintenance History	L Corrective Maintenance History	M Reliability	N Spares Lead Time	O Asset Replacement Value	P Planned Utilization	Q Visibility	R Criticality Rating	S % Maximum
4300301	10005137	CHILLER, CENTRIFUGAL, CENTRAVAC, C-1	100-B	A-100-B-MER1	$73,500	5	5	5	1	4	3	2	0	4	4	0	31	62.0
4300303	10005138	CHILLER, CENTRIFUGAL, CENTRAVAC, C-2	100-B	A-100-B-MER1	$73,500	5	5	5	1	5	1	1	0	4	4	0	28	56.0
4300311	10005292	HEAT EXCHANGER, PLATE, ALFALAVAL, HEX-1	100-B	A-100-B-MER1	$19,200	5	0	0	5	5	0	0	2	2	4	0	20	40.0
4300320	10006883	PUMP, POND WATER, PWP-1	100-B	A-100-B-MER1	$6,800	5	0	0	5	5	0	1	0	1	5	0	20	40.0
4300319	10007043	STRAINER, DUPLEX BASKET, CW, DBS-1	100-B	A-100-B-MER1	$2,675	5	0	0	2	5	0	0	2	1	5	0	20	40.0
4300344	10007017	PUMP, CONDENSER WATER, CWP-1	100-B	A-100-B-MER1	$7,925	5	0	0	4	5	0	2	0	1	3	0	18	36.0
4300351	10007018	PUMP, CONDENSER WATER, CWP-2	100-B	A-100-B-MER1	$7,925	5	0	0	1	5	0	0	0	1	3	0	11	22.0
4300352	10003796	PUMP, CHILLED WATER CHWP-1	100-B	A-100-B-MER1	$8,500	5	0	0	1	5	0	1	0	1	3	0	12	24.0
4300360	10003797	PUMP, CHILLED WATER CHWP-2	100-B	A-100-B-MER1	$8,500	5	0	0	1	4	0	0	1	1	3	0	13	26.0
4400355	10005410	AIR HANDLER, AHU-1	100-1	A-100-1-MER1	$12,900	5	0	0	5	4	0	1	0	1	3	0	13	26.0
4300378	10005411	AIR HANDLER, AHU-2	200-B	A-200-B-MER1	$10,025	5	5	5	5	4	1	2	1	2	5	0	23	46.0
4300389	10007028	AIR HANDLER, AHU-3	200-2	A-200-2-MER2	$11,350	5	0	0	5	5	0	0	1	2	5	0	20	40.0
4300334	10007029	AIR HANDLER, AHU-4	300-B	A-300-B-MER1	$9,825	5	0	0	5	4	0	0	2	2	5	0	20	40.0
4300326	10007030	BOILER, HOT WATER, PATTERSON, B-01	100-B	A-100-B-MER1	$41,750	5	5	5	5	3	1	1	1	3	4	0	33	66.0
4300396	10007031	PUMP, HEATING HOT WATER, HHWP-1	100-B	A-100-B-MER1	$6,500	5	0	0	1	5	0	1	0	0	3	0	10	20.0
4300396	10007032	PUMP, HEATING HOT WATER, HHWP-2	100-B	A-100-B-MER1	$6,500	5	0	0	1	5	0	0	0	0	3	0	11	22.0

EVALUATION CRITERIA

G   Mission Impact: Relative criticality of system or asset on the ability to meet mission or production demands
H   Safety: Could a failure of this asset or system result in a potential safety incident?
I   Environmental: Could a failure of this asset or system result in a potential reportable incident?
J   Single Point Failure: Relative value of asset that considers work-around and ability to by-pass failure in the short-term
K   Preventive Maintenance History: Average annual cost of preventive maintenance for each asset. Absence of effective PM reduces the reliability of assets
L   Corrective Maintenance History: Average annual cost for each asset. Level of expenditures is indicative of reliability
M   Reliability: From maintenance history, this is a relative ranking based on the number of breakdowns/corrective maintenance tasks required for the asset
N   Spare Parts Lead Time: Relative measure of time required to obtain spare parts or asset replacement should a failure occur.
O   Asset Replacement Value: Relative cost to replace the asset should total failure occur
P   Planned Utilization: Relative value of the planned asset utilization. Assets that are needed more than 75% are rated highest (5)
Q   Visibility: How important is the visibility impact to the general public and staff?
R   Criticality Rating: Relative number that is the sum of the eleven evaluation criteria.

Figure B-6 Equipment Criticality Assessment Worksheet (Example)

Appendix C

Articles of Interest

BEST MAINTENANCE PRACTICES

by Ricky Smith, Executive Director-Maintenance Strategies, Life Cycle Engineering®

"Best Practices." These two words have achieved a meaning of their own within the past decade. The words represent benchmarking standards for whatever area they are applied to. Nothing is better or exceeds a "Best Practice." It is the highest point towards which we measure from the lowest point. The words are most often applied to the quality of management. There exist today enormous databases of opinions (author's definition) from executives in successful companies and institutions regarding what constitutes the best business practices, the best management styles, the best corporate philosophies. Unfortunately, in some people's minds, the words "Best Practices" conjure up some obscure, always-changing and rarely achievable goal upon which they must focus with only the faintest hope of ever attaining.

Welcome back to reality. "Best Maintenance Practices" are benchmarking standards, but these are real, specific, achievable and proven standards for maintenance management that have made many maintenance departments more efficient, that have reduced facility and plant maintenance and operating costs, that have improved reliability and that have increased morale.

If everyone at your facility is satisfied with the existing maintenance program, why then, should you be interested in "Best Maintenance Practices"? Most maintenance departments in North America today operate at between 10% and 40% efficiency. Nearly 70% of facility equipment failures can be considered self-induced. These statistics can't, and shouldn't, be acceptable—not to upper management and certainly not to maintenance managers. These facts alone should generate some amount of interest in Best Maintenance Practices. Where does your maintenance department stand in relation to these figures? Do you measure and track maintenance efficiency? Do you accumulate, analyze and categorize data on equipment failures? Do you track maintenance costs for unplanned repairs or overtime? If you do none of these things, then you probably have no idea if you are the same as, better or worse than these averages.

> "The significant problems we face today cannot be solved with the same level of thinking we were at when we created them."
> —Albert Einstein

This article will introduce you to "Best Maintenance Practices," define the standards and show you the results you can expect from targeting and reaching the performance levels of Best Maintenance Practices. It will also provide you with detailed methods, strategies and actions you can put into use immediately to develop your facility's plan for implementing Best Maintenance Practices.

Best Maintenance Practices are actually defined in two separate categories. There are the standards, which are the measurable performance levels of maintenance execution; then there are the methods and strategies that must be practiced in order to meet the standards. Together, the combination of standards and methods and strategies are elements of an Integrated Planned Maintenance system. As shown in Figure 1, achieving Best Maintenance Practice Standards (classified as Maintenance Excellence), shown in gold, is accomplished through an interactive and integrated series of links with an array of processes, methods and strategies, shown in green.

Before we define the standards for Best Maintenance Practices, it may be a good idea to make sure that we all have in mind the same idea of what maintenance means:

Maintenance:
(a) to keep in its existing state
(b) preserve; continue in good operating condition; protect

"Proactive Maintenance is the Mission"

Surprisingly, there are a substantial number of people who do not know the meaning of maintenance. At least the way they practice maintenance would indicate this. In practice, the prevalent interpretation of maintenance is to "fix it when it breaks." This is a good definition for repair, but not maintenance. This style of maintenance is Reactive. As stated above, the mission is Proactive Maintenance. Here is another definition worth remembering:

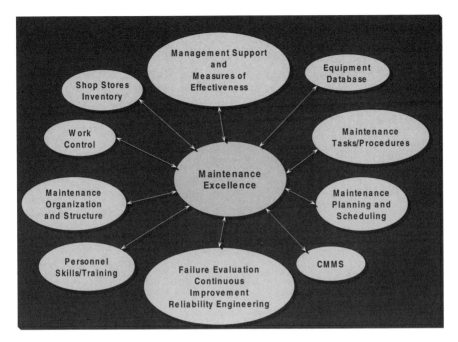

Figure C-1 Integrated Planned Maintenance System

Discipline:

(a) Self-control or orderly conduct

(b) Acceptance of or submission to authority and control; orderliness; order; control; self-control; subordination to rules of conduct, system, method

THE STANDARDS FOR BEST MAINTENANCE PRACTICES

- 100% of maintenance person's time is covered by a work order
- 90% of Work Orders are generated by Preventive Maintenance inspections
- 30% of all labor hours are from Preventive Maintenance
- 90% of planned/scheduled work compliance
- 100% reliability is reached 100% of the time
- OEE over 85%
- Spare parts stock outs are rare (less than one per month)
- Overtime is less than 2% of total maintenance time
- Maintenance budget is within +/– 2% per piece of equipment

Anyone may claim to be a maintenance expert, but the conditions within a facility/plant generally cannot often validate that this is true. In order to change the organization's basic beliefs, the reasons why an organization does not reach out to achieve these standards in the maintenance of their equipment must be identified. Two of the more common reasons that a facility does not follow best maintenance practices are: Maintenance is totally reactive and does not follow the definition of maintenance, which is to protect, preserve and prevent from decline (reactive plant culture).

The maintenance workforce lacks either the discipline to follow best maintenance practices, or management has not defined rules of conduct for best maintenance practices.

The potential cost savings of implementing Best Maintenance Practices can often be beyond the understanding or comprehension of management. Many managers are in a state of denial regarding the impact of maintenance. As a result, they do not believe that maintenance practices directly reflect on an organization's bottom line or profitability. More enlightened facilities have demonstrated that, by reducing the self-induced failures, they can increase equipment reliability by as much as 20%. Other managers accept lower reliability standards from maintenance efforts because they either do not understand the problem or they choose to ignore this issue. A good manager must be willing to admit to a maintenance problem and actively pursue a solution. How can you actively pursue a solution?

- Be Proactive, Disciplined and Accountable
- Manage to Maximize Available Resources
- Manage based on Information:
  CMMS
  Production/Operation Reports
  Feedback from Work Reports

The major emphasis for actively pursuing solutions for maintenance ineffectiveness should be on proactive thinking. Adopting a proactive approach to maintenance will improve maintenance effectiveness dramatically and more rapidly than instituting an aggressive program of maintenance effectiveness improvement within the confines of the organizational and cultural environment of an existing, predominantly reactive maintenance program.

The standards for Best Maintenance Practices at the Maintenance Management level flow down to "equipment specific" best maintenance practices that, again are benchmarks for performing preventive maintenance. Table C-1 illustrates a few of the typical equipment best maintenance practices that should become familiar, well recognized and sought after objectives of all maintenance department personnel.

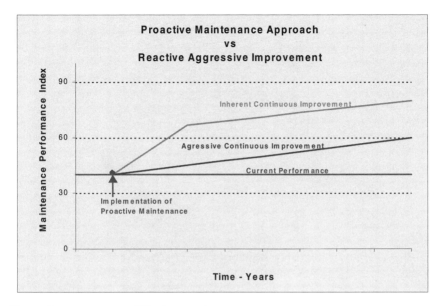

Figure C-2 Maintenance Effectiveness Improvement

Looking through this very abbreviated "Equipment Best Maintenance Practices" table, ask yourself whether your facility follows these guidelines. The results will very likely surprise you. You may find that these practices are not only not achieved in your organization, they are not even targeted as maintenance department objectives. In order to fix the problem, you must understand that the culture of the organization is at the bottom of the situation. Changing the culture is a daunting challenge; it is basic human nature to resist change. Salesmanship plays an important part in moving from a reactive to a proactive maintenance organization, which is essential if you are to succeed at the Best Maintenance Practices stratagem. There has to be a shift in mentality to allow the planning and scheduling process to work. It has been shown that when maintenance is planned and scheduled, a 25-person maintenance force operating with proactive planning and maintenance scheduling can deliver the equivalent amount of work of a maintenance crew of forty persons working in a reactive maintenance organization. Selling this concept before making the needed changes can go a long way towards easing the transition. The compelling personnel aspects of the proactive approach to maintenance include improved employee effectiveness, fewer "extended" work days, increased self-pride and the resulting improvement in employee morale.

Table C-1
Equipment Best Maintenance Practices (excerpt)

Best Maintenance Practices

Maintenance Task	Standard	Required Best Practices	Consequences of not following Best Practices	Probability of Future Failures > Number of Self Induced failures, versus following best practices
Lubricate Bearing	Lubrication interval —time based ± 10% variance	1. Clean fittings 2. Clean end of grease gun 3. Lubricate with proper amount and right type of lubricant. 4. Lubricate within variance of frequency	➤ Early bearing failure—reduced life by 20–80%.	100% ÷ 20 vs. 1

Table C-1
Continued

Coupling Alignment	Align motor couplings utilizing dial indicator or laser alignment procedures. (Laser is preferred for speed and accuracy.) Straight-edge method is unacceptable.	1. Check run out on shafts and couplings. 2. Check for soft foot. 3. Align angular 4. Align horizontal 5. Align equipment specifications not coupling specifications	➤ Premature coupling failure ➤ Premature bearing and seal failure in motor and driven unit ➤ Excessive energy loss	100% > 7 vs. 1
V-Belts	Measure the tension of v-belts through tension and deflection utilizing a belt tension gage.	1. Identify the proper tension and deflection for the belt 2. Set tension to specifications	➤ Premature belt failures through rapid belt wear or total belt failure ➤ Premature bearing failure of driven and driver unit ➤ Belt creeping or slipping causing speed variation without excessive noise ➤ Motor shaft breakage	100% > 20 vs. 1

Planning your transition for the implementation of Best Maintenance Practices is essential. Timelines, personnel assignments, documentation and all the other elements of a well-planned change must be developed before changes actually begin to take place. The following list of proactive maintenance organization attributes are the significant parts of the new approach and therefore need to be addressed in the transition plan.

* Maintenance Skills Training
* Work Flow Analysis and Required Work Flow Changes (Organizational)
* Work Order System
* Planned, Preventive Maintenance Tasks/Procedures
* Maintenance Engineering Development
* Establishment, Assignment and Training of Planner/Scheduler
* Maintenance Inventory and Purchasing Integration/Revamping
* Computerized Maintenance Management System
* Management Reporting/Performance Measurement and Tracking
* Return on Investment (ROI) Analysis
* Evaluate and Integrate use of Contractors

Maintenance Skills Training—Determine what the training is meant to accomplish. Performing a Job Task Analysis (JTA) will help you define the skill levels required of maintenance department employees. The JTA should be followed with a Skills Assessment of employee knowledge and skill levels. Analyze the gap between required skills and available skills to determine the amount and level of training necessary to close the gap. Instituting a qualification and certification program that is set up to measure skills achievement through written exams and practical skills demonstration will provide you with feedback on training effectiveness. It will also assist in resource allocation when scheduling planned/preventive maintenance tasks.

Work Flow—One element of the transition planning process that can be a major stumbling block is analyzing existing work flow patterns and devising the necessary work flow and organizational changes that must be made to accommodate your Computerized Maintenance Management System (CMMS). This process can be extremely traumatic for the employees involved, primarily because it's the nature of the beast to resist change. When work flow shifts from a reactive to a proactive posture, planned and scheduled maintenance will replace the corrective maintenance style. Your CMMS will provide insights into organized, proactive work flow arrangements through its system modeling. Although you can tailor work flow and organizational attributes to match your facility's unique requirements, it must still work within any constraints imposed by

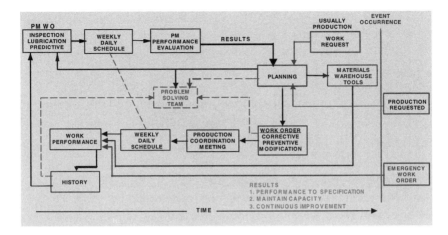

Figure C-3 Proactive Maintenance Model

CMMS Software. Of primary importance here is keeping focused on ultimate objectives—a proactive maintenance organization that will assist in reaching the standards of Best Maintenance Practices.

Work Order System—You probably have an existing work order system that is at least loosely followed. Here again, your CMMS will help you in defining changes to, or complete restructuring of, any existing work order system. The Work Order will be the backbone of the new proactive maintenance organization's work execution, information input to and feedback from your CMMS. All work must be captured on a work order—8 hours on the job equates to 8 hours on work orders. You will need to define the types of work orders your organization will need. They will include categories such as planned/scheduled, corrective, emergency, etc. The Work Order will be your primary tool for managing labor resources and measuring department effectiveness.

Planned, Preventive Maintenance Tasks/Procedures—Development of maintenance task documentation will most likely be one of the most time consuming requirements of your proactive maintenance approach, unless you already have in place the written procedures that will be used to accomplish maintenance. Procedural documentation should include standardized listings of parts, material and consumable requirements; it should identify the craft and skill level(s) required to perform the task and a frequency (or operating-time-based period) of performance. Categories of maintenance procedures that will be included in planned maintenance documentation include:

- Routine Preventive Maintenance (lube, clean, inspect, minor component replacement [e.g., filter], etc.)
- Proactive replacements (entire equipment or major component—time based or operating hours)
- Scheduled rebuilds or overhauls
- Predictive Maintenance (oil analysis, vibration monitoring, etc.)
- Condition Monitoring/Performance Based Maintenance (changes in feed speed, changes in head-flow operating point, etc.)
- Corrective Maintenance

Maintenance Engineering Development—If your facility does not have a Maintenance Engineering section, one should be established. The functions and responsibilities of new or existing maintenance engineering groups should be reviewed and revised to integrate and enhance the proactive maintenance organization. One of the alarming statistics mentioned earlier indicated that up to 70% of equipment failures are self-induced. Finding the reasons for self-induced failures, and all failures, is a responsibility of maintenance engineering. Reliability engineering is the primary role of a maintenance engineering group. Their responsibilities in this area should include evaluating preventive maintenance action effectiveness, developing predictive maintenance techniques/procedures, performing condition monitoring, planning/scheduling, conducting forensic investigations of failures including root cause analysis and performing continuous evaluation of training effectiveness.

Establishment, Assignment and Training of the Maintenance Planner/Scheduler—Whenever maintenance is performed, it is planned. It's a question of who is doing the planning, when they are doing it, to what degree and how well. Separation of planning from execution is a general rule of good management and good organizational structure. The responsibilities of the planner/scheduler are diverse, and although he or she must be familiar with the maintenance process, he or she must also be a good administrator and have the appropriate level of authority to carry out his or her role of labor usage scheduling and interfacing between many departments within the organization. The following are typical responsibilities that should be assigned to the Planner:

- Establish Equipment Numbering of all Equipment
- Develop PM Program on Each Piece of Equipment
- Ensure Accuracy of Equipment Bill of Materials
- Maintain Equipment History in CMMS as Detailed and Complete as Possible
- Review Equipment History for Trends and Recommend Improvements

- Provide Detailed Job Plan Instructions (PM Procedures)
- Determine Part Requirements for Planned Jobs
- Provide Necessary Drawings for Jobs
- Ensure Drawings are Revised and Current
- Arrange for Special Tools and Equipment
- Coordinate Equipment Downtime with Production/Operations
- Inform Production/Operations of Job Progress
- Provide Cost Information from Equipment History
- Assist with Development of Annual Overhaul Schedule
- Publish Negotiated Weekly Maintenance Schedules

The function of the Planner/Scheduler is pivotal to proactive maintenance. The position is crucial to a successful proactive maintenance approach and, therefore, vital to attaining Best Maintenance Practice standards. His assignment must be critically evaluated and he should be provided with specialized and in-depth training in his or her new role.

Maintenance Inventory and Purchasing Integration/Revamping—The cost of (parts) inventory is almost always an area where cost reduction can be substantial. With the help of suppliers and equipment vendors, purchasing can usually place contracts or Basic Order Agreements (BOA) that guarantee delivery lead time for designated inventory items. It just makes sense that your facility should shift the bulk of the cost of maintaining inventory to them.

Begin by identifying your facility's parts, material and consumable requirements. All the inventory requirements data should be entered into your CMMS. If you don't already have this data, equipment vendors can be very helpful since they usually maintain parts lists by equipment type and model. It may even be formatted such that it can be directly downloaded to your system. The parts requirements of planned/preventive maintenance tasks should then be used (your CMMS should perform this function) to generate a parts list for the planned/preventive category of work order. These are items that do not need to be in your physical inventory through the use of JIT vendor supplied BOAs.

Bar coding, continuous inventory, demand and usage data can be integrated through the use of CMMS to minimize on-hand inventory and still avoid stock outs.

Computerized Maintenance Management System—The discussion to this point has assumed that your facility has a Computerized Maintenance Management System in place. If not, or if your CMMS does not have some of the capabilities discussed here, it is certainly time to think "upgrade." An effective CMMS is critical to an organized, efficient transition to a proactive maintenance approach.

Even if your CMMS has all the capabilities needed, the transition process is an ideal time to validate the completeness and accuracy of the various CMMS module databases, particularly the equipment database. GIGO (Garbage in, garbage out) is a phenomenon that can impede or prevent you from ever achieving the standards of Best Maintenance Practices. It is also a good time to refine your work control system and to determine that the output data (CMMS report generator) is adequate to meet each user's individual requirements.

Management Reporting/Performance Measurement and Tracking— Hand-in-hand with the CMMS review and/or upgrade is the "report generator" function just mentioned. The CMMS output should be providing maintenance, engineering, production/operations, purchasing, accounting and upper management with accurate, effective and useful tools for evaluation and management. You may find that your CMMS is in need of an add-on "report generator" module or even another vendor's Report Generator software to integrate with your CMMS. At a minimum, the types of reports and data tracking you should obtain from your CMMS include:

- Open Work Order Report
- Closed Work Order Report
- Mean Time Between Failures
- "Cost per" Reports
- Schedule Compliance Report
- PM Effectiveness Report
- Labor Allocation Report
- Parts Demand/Usage Report
- Stores Efficiency Report

Return on Investment (ROI) Analysis—Justification of everything in business today is based on cost. You probably have historical data accumulated on productivity/operating costs, maintenance labor costs, maintenance material costs, inventory carrying costs and reliability/availability data; these are your key performance indicators (KPIs). Even if the data is not in these particular formats, the information for deriving this information is bound to be available. If necessary, meet with your IT group and accounting to determine how to best derive these cost figures. You will need the cost data for a minimum of the two-year period prior to beginning your transition to a proactive maintenance organization. Once you begin the planning and implementation of the changes, upgrades, etc., you will need to separate the development costs from the routine and normal operating costs of your facility to determine the total cost of implementing Best Maintenance Practices. When transition has been completed, accumulate the same cost and performance data, for your

KPIs that you obtained for the period prior to implementation. Obtaining this information must be planned for ahead of time so that you don't end up comparing apples and oranges and that you determine your real return on investment (ROI).

A.T. Kearny generated information, for the three year period following the implementation of a proactive, best maintenance practice approach in a previously reactive maintenance organization, provides the following averages:

- Productivity Increase: 28.2%
- Decrease in Maintenance Material Cost: 19.4%
- Increase in Equipment Reliability and Availability: 20.1%
- Inventory Carrying Cost Reduction: 17.8%

Depending on the size and operating costs of your facility, your realized ROI can go positive in less than three years based on typical transition costs.

Evaluate and Integrate Use of Contractors—A final item to consider when incorporating Best Maintenance Practices is integrating the use of contractors into your facility maintenance and maintenance engineering. Again, it is necessary to determine costs for in-house performance and compare them to the costs of contracting out selected efforts. This will also likely be a function of total facility size and operating costs.

Some of the maintenance or maintenance engineering efforts that may be considered as potential candidates for contractor performance include:

- Maintenance (performance of)
- Capital Improvement and/or Expansion Programs
- Predictive Maintenance (e.g., Vibration Monitoring and Analysis, Oil Analysis)
- Condition Monitoring (e.g., Equipment Performance Tests)
- Etc.

Any maintenance activities that do become a contractor function must still have relevant information/data collected and entered into your CMMS. All requirements that will be contracted to outside providers must be completely defined and should include a listing of the contractor's responsibilities and expectations prior to awarding any contracts. Formatting data for direct input to CMMS is an example of a requirement that a contractor would not routinely provide services for.

You have been introduced to "Best Maintenance Practices" (BMP) and have seen that plants and facilities today are not only failing to achieve

Best Maintenance Practice standards, but on the average, aren't even approaching acceptable maintenance effectiveness levels. You must ask yourself two fundamental questions:

- Where does my facility/plant stand relative to Best Maintenance Practices?
- Can I accept our existing maintenance effectiveness?

You must answer these questions for yourself and determine your acceptance level for performance. If you think it's time to bring you and your facility out of ineffectual practices and into cost savings, enhanced reliability and recognizable distinction, you will need to establish Best Maintenance Practices as your standards of performance. Hand-in-hand you must make a transition from a reactive maintenance organization to a totally proactive structure. The process isn't an overnight project. It will take time, effort and planning to accomplish. Above all, the transition requires commitment from all levels of your organization. The tools and planning strategies presented here will help tremendously once that commitment is made.

About the Author:
Mr. Smith is the Executive Director, Maintenance Strategies for Life Cycle Engineering. Mr. Smith is the co-author of *The Maintenance Engineering Handbook, the Plant Engineering Handbook* and *Hydraulic Fundamentals*. His contact information is: email: rsmith@LCE.com, www.LCE.com

ASSESSING YOUR MAINTENANCE TRAINING NEEDS

Ensuring That Training Fulfills Your Requirements Is Critical to Success

By Ricky Smith—Executive Director, Maintenance Strategies—Life Cycle Engineering ®

How do you know where to start with maintenance skills training? For many of us, that's the million-dollar question. That training is needed is usually self-evident. But what kind of training, in which areas, and how much training are questions not easily answered. That's what needs assessments are about.

In the Beginning

The first step in a needs assessment is to identify the problem and then determine if training will provide the answer. Many companies expect training to be the "silver bullet." But in most cases, it is only part of the real problem, which is lack of an organized and disciplined maintenance process. The diagram in Figure C-4 illustrates that many factors need to be brought together in an integrated maintenance process.

As management looks at all of these aspects of their maintenance organization, they need to find the answers to some basic questions:

- Will training resolve my problem?
- How much money will I save by implementing this training program?
- How much will the training cost?
- Is there a payback on this training?

Some hints at the answers to these questions can be found in a study funded by the U.S. Department of Education with the Bureau of Census to determine how training impacts productivity. Some of the eye-opening results were:

- Increasing an individual's educational level by 10% increases productivity by 8.6%
- Increasing an individual's work hours by 10% increases productivity by 6.0%
- Increasing capital stock by 10% increases productivity by 3.2%

Of course, training alone is not sufficient. The development and implementation of a maintenance skills training program must be part of a well-developed strategy. Skill increases that are not utilized properly will result in no changes. Once an individual is trained in a skill, he or she must be provided with the time and tools to perform this skill and must be held accountable for his or her actions.

Will Training Solve My Problem?

To answer the question, we must look into the problem. We know from research that 70% of equipment failures are self-induced; that is, equipment failures caused by the introduction of human error.

Not all self-induced equipment failures are maintenance related. Some will be induced by operator error; others, by being bumped by vehicles or other equipment, etc.

Work orders are the best source of information to determine self-induced equipment failures. We must identify the true cause of the failures through random sampling of the work orders of equipment breakdowns over a three-month period. The question to be answered: Was lack of skill the problem (self-induced failures)?

If lack of skill was the major problem, then you can easily estimate the losses due to lack of skills. First, add together the cost of production losses, the cost of maintenance labor and the cost of repair parts. Then multiply this sum by the percentage of maintenance labor hours attributable to emergency (self-induced) breakdown work orders. The final figure will be a rough indication of what your plant skills deficit is costing you.

Perform a Skills Assessment

The skill level of the maintenance personnel in most companies is well below what industry would say is acceptable. Life Cycle Engineering has assessed the skill level of thousands of maintenance personnel in the United States and Canada and found that 80% of the people assessed scored less than 50% of where they need to be in the basic technical skills to perform their jobs.

A maintenance skills assessment is a valuable tool in determining the strengths and weaknesses of a given group of employees in order to design a high-impact training program that targets those documented needs. The skills assessment should be based on the critical skills.

Maintenance personnel have often found it difficult to upgrade their technical skills because much that is available is redundant or does not take their current skill level into consideration. The assessment is designed to eliminate those problems by facilitating the construction of customized training paths for either individual of the group based upon demonstrated existing knowledge and skills.

When the assessment is used in conjunction with a job task analysis, a gap analysis can be performed to determine both what skills are needed in order to perform the job effectively and what skills the work force presently has. All training must be based on a job task analysis.

You must then fill the gaps with training that is performance based. This analysis detail identifies the exact task needed in each skill area so that all training is developed based on the actual job requirements. Gap analysis also ensures that training is EEOC compliant.

Each skill area in a skills assessment should have three components:

- Written: identifies the knowledge required for a specific skill; theories, principles, fundamentals, vocabulary and calculation should be among the skills tested.

- Identification: assesses knowledge in specific skill areas; employees are asked to name components and explain their uses in this oral assessment.
- Performance: assesses the critical skills required; to analyze this aspect, employees carry out typical maintenance tasks in accordance with generally accepted work standards.

The written assessment may be proctored by the plant's own personnel. But certified assessors from an outside agency or a local technical school should perform the identification and performance portions of the skills assessment. This practice ensures that the assessor does not have pre-conceived notions about what someone knows. Here's an example of why this precaution is important: During an assessment at a paper mill, the maintenance manager pointed to one of his employees and said, "See that man, he is the dumbest mechanic I have." The results proved otherwise. Out of 250 mechanics he rated as the fifth most skilled.

The resulting assessment data should be analyzed and compiled into a series of reports that depict scores in three ways (see accompanying charts):

- Company summary, showing a composite of all personnel tested
- Subject results, showing the scores of all personnel tested by subject area
- Individual test results, showing scores of all personnel tests by person

The results should be shared with company management as well as with the individuals tested.

The assessment report becomes a benchmark study on the status of your existing maintenance work force and is useful as the tool against which to measure progress or as the profile against which to hire new employees in order to round out the department.

After completion of the assessment process, you can begin work to establish performance standards for each employee or for the group, develop a training plan to address the identified needs, develop curriculum to meet those training goals or deliver training in the targeted skills.

Increasing pressure to improve productivity and reduce costs is forcing organizations to search for innovative solutions. Targeted training is both effective and efficient, regardless of whether the goal is to design a full apprentice-to-journeyman program or just identify skills for high-impact brushing up.

Time and money spent on a training needs assessment will help you get the most out of the limited training dollars available by helping identify the training opportunities allowing money to be allocated effectively.

Facing the Facts About Maintenance Skills

- Most companies do not have fully skilled maintenance personnel
- You cannot fire everyone that is incompetent
- Hiring skilled maintenance personnel is difficult
- Most repetitious equipment problems that cost companies billions of dollars a year are a direct result of skill deficiencies
- A person that feels competent is a better worker and is motivated easier
- Often maintenance personnel are disciplined because of skill deficiency, not because of a lack of concern or commitment
- People become frustrated or stressed when they do not know the proper way to do a specific task
- Companies spend millions of dollars a year on maintenance training without regard to the results expected from it or without a way of measuring results (money spent does not always equal value received)

While skills training is important, it is only one factor among the many that make a successful maintenance operation.

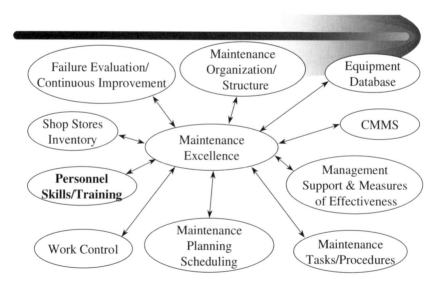

Figure C-4 The Integrated Maintenance Process

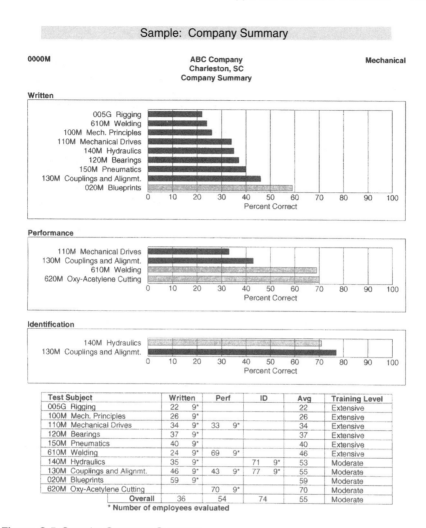

Figure C-5 Sample: Company Summary

Data gathered in the skills assessment process should be analyzed in three ways: by each skill for the entire plant or company (company summary), overall skill level for each employee (skill summary) and by each skill for each individual (individual summary). See Figure C-5.

Sample: Skill Summary

0000M

ABC Company
Charleston, SC
130M Couplings and Alignmt.

Mechanical

Written

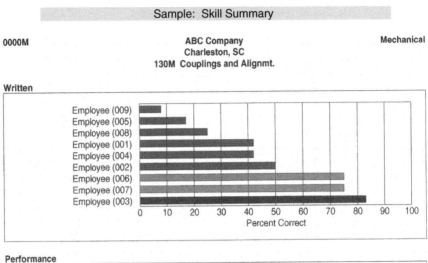

Performance

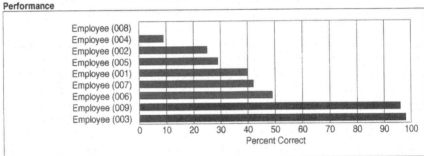

Identification

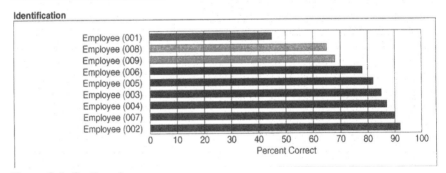

Figure C-5 *Continued*

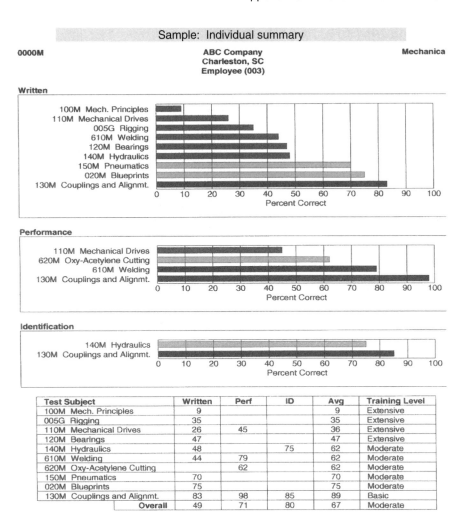

Test Subject	Written	Perf	ID	Avg	Training Level
100M Mech. Principles	9			9	Extensive
005G Rigging	35			35	Extensive
110M Mechanical Drives	26	45		36	Extensive
120M Bearings	47			47	Extensive
140M Hydraulics	48		75	62	Moderate
610M Welding	44	79		62	Moderate
620M Oxy-Acetylene Cutting		62		62	Moderate
150M Pneumatics	70			70	Moderate
020M Blueprints	75			75	Moderate
130M Couplings and Alignmt.	83	98	85	89	Basic
Overall	49	71	80	67	Moderate

Figure C-5 *Continued*

A CMMS HIDDEN TREASURE—HISTORY

By Scott Franklin, Vice President, Life Cycle Engineering®

Investing in a CMMS is very similar to any other capital investment. You prepare a budget (oftentimes supported by a return on investment calculation), perform a selection process, make the purchase, perform the installation and begin operations. One significant difference between the CMMS investment and other capital investments is that, over time, your return on investment for a CMMS can actually increase. This is because the CMMS system is doing something hardware doesn't. It is collecting information that can then be used to further improve the efficiency of your processes.

Installing a CMMS is generally viewed as a two-stage process. The first stage being everything necessary to make the system operational followed by the second stage of operating the system. However, there is a hidden third stage. As the system is used to purchase repair parts, issue and track work orders and schedule/plan maintenance and capital projects, a wealth of history is being collected. Since there is no quantifiable point where historical data automatically becomes useful (at least not in the same manner that "going live" with the system is), it is easy to overlook the fact that over time, you have collected a significant amount of information. This hidden third stage is the ability to analyze this collected information to improve and optimize your maintenance program.

In reality there is a point where your historical information has 'gone live' and this is when you have one calendar (or fiscal) year's worth of information. Let's step back and make a few assumptions. One being that the Thanksgiving/Christmas holiday spirit is usually greatly enhanced with the always-enjoyable Annual Budget Preparation festivities. The second assumption is that you also get to enjoy starting the new year spending some quality time with your boss reviewing last year's maintenance expenses (readily available at your local accounting department). Simply put, every year provides two discrete data points—an itemized cost projection/budget and an associated year-end actual cost totals. If your CMMS system has been in operation for that entire year, then you now have a third piece of information, your CMMS totals. However, one small problem may exist. Chances are, your CMMS totals don't match anything.

The reason for the mismatch is that your accounting process is basically designed to track expenses and your CMMS is designed to track equipment, labor and parts. This does not mean that they are mutually exclusive, but that they are optimized to track common information differently. Ideally, the CMMS should be configured to collect and categorize the same cost information that your budgeting/accounting process does. For example, how is your budget categorized? Capital improvements versus general maintenance repairs, corrective maintenance versus preventive maintenance, etc., and is your CMMS system set up to match these categories? Assuming you have a year's worth of data available, try printing reports to match the original budget estimates. If there are certain areas of your budget that you don't track within your CMMS system—such as capital projects or equipment upgrades—then you may want to consider using the CMMS for these items also. Most CMMS systems have matured to the point that the most recent releases have significantly expanded their capabilities beyond the basic maintenance management/equipment history/purchasing/inventory functionality of the original products. This means that the CMMS you purchased 4–5 years ago has probably had one major upgrade and numerous minor upgrades and may include significantly greater functionality and features that you may not be aware of.

In addition to annual budgets, there is also the availability of cost information from your accounting system. If your accounting system is not integrated with your CMMS system, then getting accurate correlation between Accounting's Accounts Payable entries and CMMS parts/equipment costs can be difficult. You may need to work with your accounting department to find some way of recording a common identification number between the two systems—e.g., recording the Purchase Request/Purchase Order number, the invoice number or the work order number. Ideally, all parts/equipment are purchased against a work order and coordinating with your purchasing department to record the Work Order Number can greatly simplify resolving discrepancies between CMMS numbers and accounting numbers. It should also be possible to identify or develop a report from the accounting system that can itemize costs by Work Order number and allow direct comparisons with the CMMS system.

The most obvious advantage to configuring your CMMS to match your budgeting/accounting process is that this information is now readily available in the CMMS and greatly simplifies the budget preparation and adds near-real time cost tracking. Additionally, the ability to show figures from the CMMS system that directly correlate to budgeting/accounting numbers creates inherent credibility when presenting information from the CMMS.

While there are definite advantages to having budgets, accounting costs and CMMS totals correlate, this may not always be possible and/or cost effective. Configuring your CMMS to collect and categorize costs to match your budgeting/accounting procedures is a useful guideline to ensure quantifiable costs are being recorded. The real value in this historical information is how a review of history can improve future operations. For example:

1. Expense Analysis: There are a number of areas that can be analyzed by expenses. These include:
   - Corrective versus Preventive costs: Realizing that preventive maintenance cost are primarily labor costs and consumables (with the notable exception of scheduled overhauls/replacements), building a reviewable history of the relationship between Preventive Maintenance and Corrective Maintenance costs allows the potential to build a "Return on Investment" of your Preventive Maintenance program, especially if you can track associated/ estimated costs for unscheduled downtime/equipment failures
   - Capital Improvements versus Maintenance/Repair costs: Capital expenditures/equipment replacements (especially unplanned replacements) can often skew maintenance costs analysis; tracking these in the CMMS can simplify the analysis of actual maintenance and repair costs.
2. Cost Tracking: Tracking maintenance expenses/costs simplifies the ability to stay within budgets and also justify next year's budget requests. Properly recording and categorizing maintenance expenses allows on-demand reporting of year-to-date costs and comparison to budget projections.
3. Equipment Cost Analysis: A recent review of one company's CMMS showed that they were spending $17,000 per year to maintain a group of pumps that had a $350 replacement cost per pump. This had been going on for a number of years and wasn't immediately obvious until:
   a) A year's worth of information had been collected and,
   b) That information was reviewed.

   Most CMMSs include a "Top Ten" or "Top Twenty" listing based on various criteria (maintenance cost in this example) that can be used to identify trends. Be aware that you may have to look a little deeper to identify some problems. The problem cited above wasn't noticed until the report was run by grouping all identical equipment (i.e., a cumulative total for all the $350 pumps).
4. Equipment Reliability: A simple comparison of preventive maintenance versus corrective maintenance can be quite revealing. A make/model of equipment that is 25% less expensive than a similar model by a different manufacturer, but has double the maintenance problems may not be your best investment. A review of history can also identify areas where you may want to reevaluate your maintenance focus. Equipment with a high amount of corrective maintenance is a prime candidate for reviewing the quantity and quality of your preventive maintenance. Conversely, equipment with a high

amount of preventive maintenance, but very little corrective maintenance may not need quite so much attention.

5. Labor Analysis: If you have been the Maintenance Department director for a number of years, chances are you have been faced with the requirement to justify your staffing levels or justify increasing your manpower. Additionally, the availability of quantitative data would greatly simplify the justification process. By using the CMMS system to document labor hours, over time you can build a significant amount of information. The primary benchmark is how much time is being documented. A reasonable target is to try to get anywhere from 50% to 80% of the labor hours documented in the CMMS system—i.e., labor hours documented against a work order. Being able to show a relatively high utilization along with a comparative breakdown of preventive maintenance hours versus corrective maintenance/capital improvements can provide significant ammunition when faced with labor justification questions. (Note— Ensure that you have a defensible explanation for the utilization percentage. Your maintenance supervisors should be able to give you a realistic target number and a justification for that number. Make sure you take the time to phrase the question constructively).

These are just some examples of different analysis possibilities. The real "secret" is to recognize that information exists in your CMMS that wasn't there the day you started using the system. Take some time to see what you have. It may surprise you.

Scott Franklin is a Vice President at Life Cycle Engineering, Inc.— www.lce.com—in Charleston, S.C.

TRENDS IN MAINTENANCE

A Survey of Maintenance Practices

There is no question that the maintenance function is receiving more and more attention today. Plant and facility managers are realizing the impact that maintenance and its effect on equipment reliability can have on operating costs, productivity and profit.

In 1995, Life Cycle Engineering, Inc. (LCE) conducted a survey of maintenance practices in order to determine where North American maintenance organizations stood in relation to Best Maintenance Practices (BMP). Over the last few months of 2002, another, more abbreviated, survey was conducted to obtain a thumbnail picture of any progress made towards an improved maintenance function. The survey results are provided here and compared with previous results to gauge any progress.

The Survey

The 2002 Maintenance Practices Survey consisted of four questions addressing areas of maintenance that are most indicative of how effective an organization is in maintaining a high level of equipment reliability and how well organized and disciplined the maintenance function is. On average, each question received 158 responses. These were grouped by survey response category and are shown as a percentage of total responses.

Survey Question 1 (Figure C-6): What percentage of total maintenance labor hours by week/month/year are spent performing preventive maintenance tasks?

Survey Question 2 (Figure C-7): What is the ratio of preventive maintenance (PM) work orders to corrective maintenance (CM) work orders obtained by Preventive and Predictive Maintenance (PM/PdM) inspections?

Survey Question 3 (Figure C-8): What is the ratio of PM labor hours to Emergency labor hours?

Survey Question 4 (Figure C-9): What percentage of PM compliance, based on the scheduled frequency, do you meet? (i.e., 10% means that a 7-day PM is completed within 7 hours of schedule)

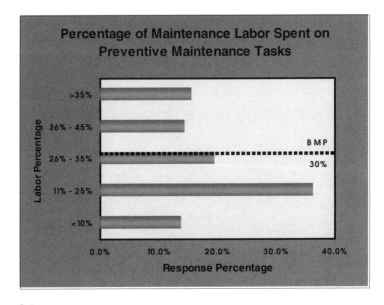

Figure C-6

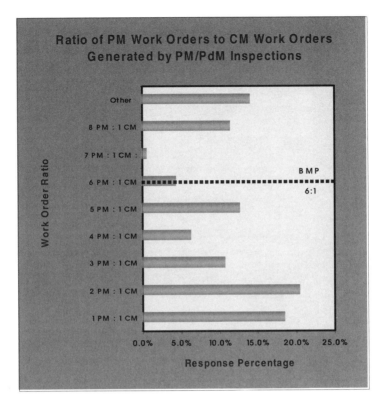

Figure C-7

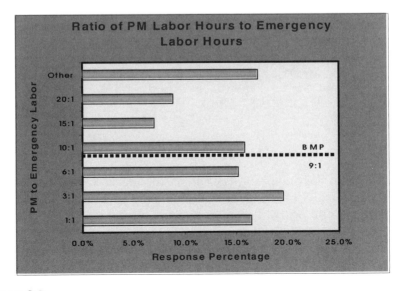

Figure C-8

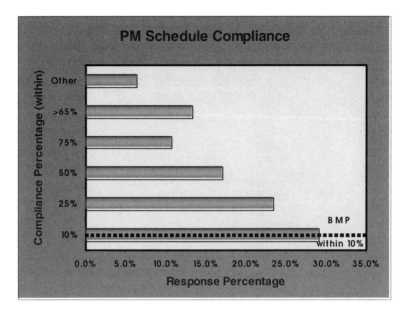

Figure C-9

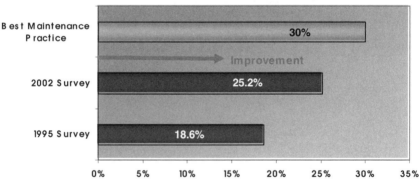

Figure C-10

The Trend

In the following graphical illustrations, the results of this survey are compared with the 1995 survey as well as with the Best Maintenance Practice criteria for that particular maintenance practice.

NOTE:

The 6:1 BMP ratio implies that for every six times a piece of equipment is inspected one inspection will result in generating a corrective maintenance work order. The ratio is based on "acceptable risk." This ratio can, and should improve even further if RCM Analysis is being performed.

Ratio of PM Work Orders to PM-Generated CM Work Orders

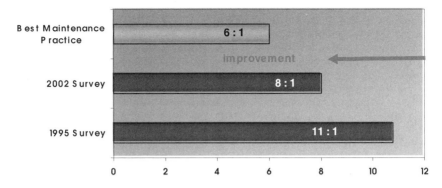

Figure C-11

Ratio of PM Labor to Emergency Labor

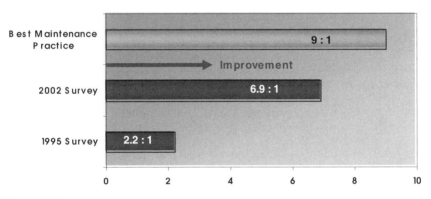

Figure C-12

Percentage of PM Compliance

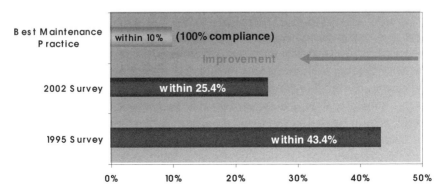

Figure C-13

NOTE: The 9:1 BMP ratio implies that 30% of all maintenance labor should be spent doing PM while only 2–5% of all maintenance labor should be spent doing emergency work. The percentage is based on "acceptable risk." This ratio can, and should improve even further if RCM Analysis is being performed.

The previous graphic illustrations show clearly that there has been significant improvement in maintenance practices and maintenance effectiveness since 1995. In large part, this has been due to a growing acceptance and implementation of Computerized Maintenance Management Systems (CMMS) and the resulting tighter integration of plant and facility maintenance, operations/production and purchasing/stores organizations. There is still room for continued improvement and progress towards achieving the standards for Best Maintenance Practices.

Bear in mind that the numbers used for comparison of the 1995 and 2002 surveys are averages. Some organizations have achieved the standards of Best Maintenance Practices while others have a very long way to go. Between these extremes are a significant number of plants and facilities showing progress, but perhaps thwarted or slowed by factors such as inadequate planning, lack of management support, lack of capital expenditure commitments or other impediments.

Planning for the implementation of Best Maintenance Practices is essential. Timelines, personnel assignments, documentation and all the other elements of a well-planned change must be developed before real change begins to take place. The following list of proactive maintenance organization attributes are significant parts of the approach to achieving Best Maintenance Practices and therefore need to be addressed in the transition plan.

- Top Management Commitment
- Maintenance Skills Training
- Work Flow Analysis and Required Changes (Organizational)
- Work Order System
- Planned, Preventive Maintenance Tasks/Procedures
- Maintenance Engineering Development
- Establishment, Assignment and Training of Planner/Scheduler
- Maintenance Inventory and Purchasing Integration/Revamping
- Computerized Maintenance Management System (CMMS)
- Management Reporting/Performance Measurement and Tracking
- Return on Investment (ROI) Analysis
- Evaluate and Integrate Use of Contractors

Within a few years, the trend of Maintenance Function improvement can show averages that are in line with Best Maintenance Practices. Maintenance Managers have a serious responsibility for keeping this trend as active as possible. This kind of continued improvement is essential if production plants in North America are to be competitive in the global economies of today and tomorrow.

KEY PERFORMANCE INDICATORS

Leading or Lagging and When to Use Them

By Ricky Smith, Executive Director-Maintenance Strategies, Life Cycle Engineering®

Initiating major change, such as moving from a reactive maintenance operation to one, which is proactive and employs Best Maintenance Practices to achieve Maintenance Excellence, requires start-up support from top management. In order to continue the journey towards Maintenance Excellence, the continued support from management will need justification. Upper management will not be satisfied with statements like "just wait until next year when you see all the benefits of this effort." They will want something a little more tangible if you are to gain further commitment from them. You will need to provide tangible evidence in the form of objective performance facts.

That's where metrics comes in. *Metrics* is just a term meaning "to measure" (either a process or a result). Combining several metrics yields indicators, which serve to highlight some condition or highlight a question that we need an answer to. Key Performance Indicators (KPI) combine several metrics and indicators to yield objective performance facts. They provide an assessment of critical parameters or key processes. KPI for maintenance effectiveness have been discussed, defined and refined for as long as proactive maintenance has been around. KPIs combine key metrics and indicators to measure maintenance performance in many areas.

Metrics can be a double-edged sword. Metrics are essential for establishing goals and measuring performance. Metrics chosen or combined erroneously can produce misleading indicators that yield incorrect and/or low performance measures. Inaccurate measures produce bad management decisions.

If you are involved in an equipment improvement program, such as Maintenance Excellence, you must have a thorough understanding of the financial metrics used by your company to measure results and track improvement. You will need to establish a direct link between improved equipment reliability and overall company operational performance. At the bottom line, your metrics must yield a KPI in terms of financial performance.

To determine maintenance strengths and weaknesses, KPI should be broken down into those areas for which you need to know the performance levels. In maintenance these are areas such as preventive maintenance, materials management process, planning and scheduling and so on until two major Maintenance Department KPIs are defined:

- Maintenance Department Operating Costs (Budget Performance)
- Equipment Reliability

In turn, equipment reliability must correlate to production—both production versus capacity and cost per unit produced. On the other hand, operating costs must be carefully considered. Initiating change is going to initially increase maintenance department expenses. Accurately forecasting a budget centered on change is essential if KPI is going to accurately depict department budget performance (see Figure C-14).

Depending on KPI values, we classify them as either leading or lagging indicators. Leading indicators are metrics that are task specific. They respond faster than results metrics and are selected to indicate progress towards long-term objectives. Leading indicators are indicators that measure and track performance before a problem arises. To illustrate this, think of key performance indicators as yourself driving a car down a road. As you drive, you deviate from the driving lane and veer onto the shoulder of the road. The tires are running over the "out of lane" indicators (typically a rough or "corrugated" section of pavement at the side of the road that serves to alert you to return to the driving lane before you veer completely off the pavement onto the shoulder of the road). These "out of lane" indicators are the KPI that you are approaching a critical condition or problem. Your action is to correct your steering to bring your car back into the driving lane before you go off the road (proactive condition).

If you did not have the indicators on the pavement edge, you would not be alerted to the impending crisis and you could veer so far out of the driving lane that you end up in the ditch. The condition of your car, sharply listing on the slope of the ditch, is a lagging indicator. Now you must call a wrecker to get you out of the ditch (reactive condition). Lagging indicators such as your budget, yield reliability issues, which will result in capacity issues.

The necessity for tracking KPIs other than just Equipment Reliability and Budget Performance is to pinpoint areas responsible for negative trends (leading indicators). You would not want to scrap your Maintenance Excellence initiative when the only problem is that the Planner/ Scheduler didn't receive adequate training. By observing and tracking

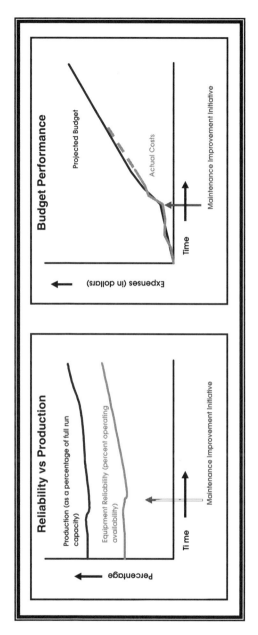

Figure C-14 Measure of Budget Performance

Planned/Schedule Compliance and Planned Work as a percentage of total labor you should be able to detect "non-improving" or even negative performance early enough to identify and correct the training problem. The "lower tier" leading indicators are also necessary for establishing benchmarks (Best Maintenance Practices) and tracking departmental progress. For example, the benchmark for the KPI "Planned/Schedule Compliance" is generally accepted as 90%. The tracking and public display of positive leading KPIs also provide significant motivational stimuli for maintenance department personnel.

A manager must know if his department is squarely in the driving lane and that everything is under control as long as possible before it approaches and goes into the ditch. A list of some of the key performance indicators of the leading variety are illustrated in Table C-2. Note that some of these indicators could be both leading and lagging when combined with and applied to other KPIs (Key Performance Indicators).

MRO STOREROOM PRACTICES—BEST OR WORST

By Ricky Smith, Executive Director-Maintenance Strategies, Life Cycle Engineering®

MRO Storeroom practices can be the root cause of a significant number of equipment reliability problems (client surveys indicate a range of 25% to 75%) in a maintenance operation. It is important to understand that this situation is not created by lack of a dedicated storeroom staff nor an incompetent storeroom staff. The situation generally results from a lack of a maintenance support strategy, the lack of an effective operating process and a lack of management focus.

What are the risks associated with MRO (maintenance, repair and overhaul) Storeroom practices? Of course there are the obvious—stock outs of critical spares, wrong parts delivered and stored under valid part numbers, etc. But what about the not so obvious risks—risks of the unknown? These are perhaps the most insidious simply because they are unknown. You don't know where they are and you don't know what your level of risk is.

What are some of the most common unknown risk factors associated with storeroom practices? The following examples have the potential for significant risk:

- Used Parts can and do represent a high rate of failure because the reliability life is unknown. These parts could last from two minutes to two years. Can you accept this risk?

Table C-2
Key Performance Indicators

Reliability/Maintainability

➢ MTBF (mean time between failures) by total operation and by area and then by equipment.

➢ MTTR (mean time to repair) maintainability of individual equipment.

➢ MTBR (mean time between repairs) equals MTBF minus MTTR.

➢ OEE (overall equipment effectiveness) Availability × Efficiency (slow speed) × Quality (all as a percentage)

Preventive Maintenance (includes predictive maintenance)

➢ PPM labor hrs. divided by Emergency labor hrs.

➢ PPM WOs (work orders) #s divided by CM (corrective maintenance, planned/scheduled work) WOs as a result of PM inspections.

Planning and Scheduling

➢ Planned/Schedule Compliance—(all maintenance labor hours for all work must be covered and not by "blanket work orders") this a percentage of all labor hours actually completed to schedule divided by the total maintenance labor hours.

➢ Planned work—a % of total labor hours planned divided by total labor hours in scheduled.

Materials Management

➢ Stores Service Level (% of stock outs)—Times a person comes to check out a part and receives a stock part divided by the number of times a person comes to the storeroom to check out a stocked part and the part not available.

➢ Inventory Accuracy as a percentage.

Skills Training (NOTE: A manager must notify maintenance craft personnel about the measurement of success of skills training

➢ MTBF.

➢ Parts Usage—this is based on a specific area of training such as bearings.

Maintenance Supervision

➢ Maintenance Control—a % of unplanned labor hours divided by total labor hours.

➢ Crew efficiency—a % of the actual hours completed on scheduled work divided by the estimated time.

➢ Work Order (WO) Discipline—the % of labor accounted for on WOs.

Work Process Productivity

➢ Maintenance costs divided by net asset value.

➢ Total cost per unit produced.

➢ Overtime hours as % of total labor hours.

NOTE: KPIs must answer questions that you as a manager ask in order to control your maintenance process. Listed is a sampling of recommended KPIs. They are listed by the areas in which a maintenance manager must ask questions.

- Parts not "Installation Ready" for equipment repair, e.g., motors with rusted or nicked shafts, which could cause a maintenance person, pressed to get equipment back on-line, to install a coupling with a hammer resulting in premature failure of the motor.
- No PM coverage on appropriate storeroom parts such as large gearboxes, motors and large bearings. PMs may be as simple as inspection of protective packaging or preservation measures.
- Unknown Shelf Life and/or Storage Specifications can also cause premature failures, for example: storing electrical components in a dusty or humid environment or using conveyor belting whose shelf life is 12 months and is checked out after 36 months.
- Parts used, but not logged or drawn on a work order create an unknown that only becomes apparent the next time that part is needed and is not available because nothing generated a reorder.
- The unrestricted use of consumables (e.g., lubricants) creates an unknown through lack of usage data and failure to trigger low level auto-reorder.
- Ineffective parts locator system creates lost time spent hunting throughout the storeroom for a part. This can represent an unknown resulting in production line downtime while waiting for a critical part. It could also result in using a "close enough" part, thus changing the specifications and reliability of a piece of equipment.

THE FIRST STEP

The solutions for eliminating these kinds of risk factors are elimination of the "unknown" characteristic of the factor. Here are some tips for incorporation into your storeroom practices:

- Identify and track the storage requirements and shelf life limits on applicable spares.
- Have vendors who rebuild motors, gearboxes, etc., return the item in a "ready to install" state; perform receipt inspections.
- Extend your Preventive Maintenance program to specified spares, which need attention at defined intervals; ask your vendors for PM procedures and frequency.
- Update (continuously) plant equipment inventory lists and associated repair parts listings including the purging of removed items.
- Identify/associate all storeroom inventory with equipment application (based on the updated plant equipment inventory list).
- Establish a "parts for work order process" (no part issued without a work order).

- Establish a bar code system for the storeroom (identifies part, equipment application and storeroom location).
- Establish and enforce the use of a log sheet for all "after hours" parts and consumables usage.
- Facilitate consolidation of parts, establishing proper stocking levels, assignment of minimum and maximum levels for each item.
- Work with maintenance to eliminate "private parts-stockpiles"; their condition is unknown and they generate no usage data, eventually resulting in stock outs.
- Track such things as premature failure of equipment/parts taken from stores inventory (less than 0.1%) and stock outs (less than 2%) in the storeroom; perform cycle counts at a set time interval (inventory accuracy of 98% or better).
- Visit all rebuild operations to ensure they have your interest in mind.

For their mutual success, Maintenance Management and Stores Management must work together because they each help define storeroom practices. Maintenance Management must provide a clear direction to the maintenance staff for assistance in eliminating stores problems. The bottom line is that all parts that leave the storeroom must be able to provide maximum reliability, not provide equipment with unknown and unacceptable reliability risks.

THE NEXT STEP

When your plant's operation matures in overall efficiency, there is one single step that MRO Storerooms can take to eliminate almost all of these risk factors of unknown magnitude: shift as much of the stores risk burden to your suppliers as practical.

How can you shift responsibility outside of the plant? Actually it's pretty simple and it's all based on JIT (just-in-time) supply management. To be successful though, your plant must have a CMMS (Computer Managed Maintenance System) installed, implemented and fully integrated between Maintenance, Purchasing and the MRO Storeroom.

In the past, one of the objectives of the maintenance storeroom was to have on hand as many spare parts as possible, just in case they might be required. But the increasing cost of inventory is making that practice obsolete. Over half of existing inventory can be eliminated by using your CMMS to schedule, based on maintenance action scheduling, when repair parts and consumables will be needed. Then your supplier is provided with specific delivery requirements to meet those maintenance schedules

(JIT delivery). All PM-related parts and consumables can be removed from your storeroom. Incoming items can go directly to a staging area (much smaller than the storeroom space previously occupied) for issue to the following week's PM work orders.

Accurate equipment inventories in your CMMS aid in accurately identifying repair parts requirements. By providing these requirements to your supplier, you can contract for minimum lead time delivery (JIT) of those parts. Now instead of carrying sufficient parts for six month's usage, minimum lead times (for example—one week) will allow you to reduce that level to one week's usage. You've just achieved a 26:1 reduction in on-hand parts inventory.

CMMS can also supply you with multiple application data for repair parts. If a part is used in six different equipments, you don't need to keep six of them in your spares inventory. Two would be more than ample, and with minimum lead time (JIT) ordering, you could probably reduce that item in inventory to one. A reduction here of as much as 6:1.

Predictive Maintenance (PdM) measures equipment condition related data (vibration levels, temperature, oil viscosity, speed, etc.), which can be trended by your CMMS to predict when equipment operating specifications will be exceeded. These predictions can then be used to trigger repair parts ordering based on procurement lead time and predicted out-of-tolerance condition (JIT delivery).

CMMS can generate parts usage data, which should be used to reorganize your storeroom. High-usage items should be quickly accessible and low usage items can go to the "back of the storeroom." A word of caution: all parts locations—high or low usage—must be accurately identified. Accessibility means the time it takes to walk directly to the part's location, not how long it takes to hunt the part down. CMMS is also used to enter and retrieve parts location information (which can be greatly enhanced through the use of bar coding).

A final few words of wisdom are appropriate at this point. The previous paragraphs have discussed the utilization of CMMS to reduce storeroom inventory and increase storeroom efficiency. Having CMMS does not automatically accomplish these efficiencies, rather CMMS is a prerequisite to MRO Storerooms implementing these processes. It is the practice, not the CMMS that facilitates inventory reduction and improved storeroom efficiency.

About the Author:
Ricky Smith, CMRP, CPMM, is the Executive Director of Maintenance Solutions for Life Cycle Engineering, Inc.
rsmith@LCE.com, www.LCE.com, 843-744-7110 ext. 350

WHAT IS LEAN MAINTENANCE?

By Ricky Smith, Executive Director-Maintenance Strategies, Life Cycle Engineering®

Much has been written about Lean Manufacturing and the Lean Enterprise; enough that nearly all readers are familiar with the concepts as well as the phrases themselves. But what about Lean Maintenance? Is it merely a subset of Lean Manufacturing? Is it a natural "fall-in-behind" spin-off result of adopting Lean Manufacturing practices? Much to the chagrin of many manufacturing companies, whose attempts at implementing Lean practices have failed ignominiously, Lean Maintenance is neither a subset nor a spin-off of Lean Manufacturing. It is instead a prerequisite for success as a Lean manufacturer. Why is that?

Perhaps the best starting point is to define Lean Maintenance:

> Lean maintenance is a proactive maintenance operation employing planned and scheduled maintenance activities through Total Productive Maintenance (TPM) practices using maintenance strategies developed through application of Reliability Centered Maintenance (RCM) decision logic and practiced by empowered (self-directed) action teams using the 5-S process, weekly Kaizen improvement events, and autonomous maintenance together with multi-skilled, maintenance-technician-performed maintenance through the committed use of their work order system and their Computer Managed Maintenance System (CMMS) or Enterprise Asset Management (EAM) system supported by a distributed, Lean maintenance/MRO storeroom that provides parts and materials on a just-in-time (JIT) basis and backed by a maintenance and reliability engineering group that performs root cause failure analysis (RCFA), failed part analysis, maintenance procedure effectiveness analysis, predictive maintenance (PdM) analysis and trending and analysis of condition monitoring results.

That's it in a nutshell, albeit a rather large nut (except for a few details that were omitted here but will be covered later). Maybe it would be a good idea to discuss the highpoints of this definition just to be sure everyone understands all of the terms used.

- **Proactive**—The opposite of reactive wherein the maintenance operation reacts to equipment failures by performing repairs. In the proactive maintenance operation, the prevention of equipment failures through performance of preventive and predictive maintenance actions is the objective. Repair is not equivalent to maintenance.
- **Planned and Scheduled**—Planned maintenance involves the use of documented maintenance tasks that identify task action steps, labor resource requirements, parts and materials requirements, time to perform and technical references. Scheduled maintenance is the prioritization of the work, issuance of a work order, assignment of available labor resources, designation of the time period to perform the task (coordinated with operations/production) and breakout and staging of parts and materials.
- **Total Productive Maintenance (TPM)**—The very foundation of Lean Maintenance is Total Productive Maintenance (TPM). (Refer to the Maintenance Management Pyramid.) TPM is an initiative for optimizing the reliability and effectiveness of manufacturing equipment. TPM is team-based, proactive maintenance and involves every level and function in the organization, from top executives to the shop floor. TPM addresses the entire production system life cycle and builds a solid, shop-floor-based system to prevent all losses. TPM objectives include the elimination of all accidents, defects and breakdowns.
- **Reliability Centered Maintenance (RCM)**—A process used to determine the maintenance requirements of physical assets in their present operating context. While TPM objectives focus on maintaining equipment reliability and effectiveness, RCM focuses on optimizing maintenance effectiveness.
- **Empowered (Self-Directed) Action Teams**—Action Team activities are task oriented and designed with a strong performance focus. The team is organized to perform whole and integrated tasks hence requiring multi-department membership. The team should have defined autonomy (that is, control over many of its own administrative functions such as self-evaluation and self-regulation all with limits defined.) Furthermore, members should participate in the selection of new team members. Multiple skills are valued. This encourages people to adapt to planned changes or occurrence of unanticipated events.
- **5-S Process**—Five activities for improving the workplace environment: 1. Sort (remove unnecessary items), 2. Straighten (organize), 3. Scrub (clean everything), 4. Standardize (standard routine to sort, straighten and scrub) and 5. Spread (expand the process to other areas).

- **Kaizen Improvement Events**—Kaizen (a Japanese word) is the philosophy of continuous improvement, that every process can and should be continually evaluated and improved in terms of time required, resources used, resultant quality and other aspects relevant to the process. These events are often referred to as a Kaizen Blitz—a fast turn-around (one week or less) application of Kaizen "improvement" tools to realize quick results.

- **Autonomous Maintenance**—Routine maintenance (e.g., equipment cleaning, lubrication, etc.) performed by the production line operator. The Maintenance Manager and Production Manager will need to agree on and establish policy for (1) Where in the production processes autonomous maintenance will be performed, (2) What level and types of maintenance the operators will perform and (3) How the work process for autonomous maintenance will flow. Specific training in the performance of designated maintenance responsibilities must be provided to the operators prior to assigning them autonomous maintenance responsibilities.

- **Multi-Skilled Maintenance Technician**—Multi-skilled maintenance technicians are becoming more and more valuable in modern manufacturing plants employing PLCs, PC Based Equipment and Process Control, Automated Testing, Remote Process Monitoring and Control and/or similar modern production systems. Maintenance Technicians who can test and operate these systems as well as make mechanical and electrical adjustments, calibrations and parts replacement obviate the need for multiple crafts in many maintenance tasks. The plant processes should determine the need for and advantages of including multiple skills training in the overall training plan.

- **Work Order System**—The system used to plan, assign and schedule all maintenance work and to acquire equipment performance and reliability data for development of equipment histories. The work order is the backbone of a proactive maintenance organization's work execution, information input, and feedback from the CMMS. All work must be captured on a work order—8 hours on the job equals 8 hours on work orders. The types of work orders will include categories such as planned/scheduled, corrective, emergency, etc. The work order will be the primary tool for managing labor resources and measuring department effectiveness.

- **Computer Managed Maintenance System (CMMS)**—An information (maintenance) management software system that performs, as a minimum, Work Order Management, Planning Function, Scheduling Function, Equipment History Accumulation, Budget/Cost Function, Labor Resource Management, Spares Management and a

Reports Function that utilizes Key Performance Indicators (KPI). To be effective, CMMS must be fully implemented with complete and accurate equipment data, parts and materials data and maintenance plans and procedures.

- **Enterprise Asset Management (EAM)**—EAM Systems perform the same functions that CMMS does but on a more organization-wide, integrated basis, incorporating all sites and assets of a corporation. Even broader Enterprise Systems (ES) incorporate fully integrated modules for all the major processes in the entire organization and offers the promise to effectively integrate all the information flows in the organization.

- **Distributed, Lean Maintenance/MRO Storeroom**—Several stores locations replace the centralized storeroom in order to place area specific parts and materials closer to their point-of-use. Lean stores employ standardized materials for common application usage. The Lean stores operation also employs planning and forecasting techniques to stabilize the purchasing and storeroom management process. This method requires that a long-term equipment plan is developed and equipment Bills of Material (BOM) are entered into the CMMS system as soon as the Purchase Order for new equipment is issued.

- **Parts and Materials on a Just-In-Time (JIT) Basis**—Stores inventories are drastically reduced (as are the costs of carrying large inventories) through a strong supply chain management team that uses just-in-time suppliers, and practices such as vendor-managed inventories (VMI) in which the vendor is given the responsibility for maintaining good inventory practices in replenishment, in ordering and issuing the materials. The vendor is charged with the responsibility of controlling costs, inventory levels, the sharing of information with the facility and improvements in the process. The supply chain management team advocates day-to-day supplier communication and cooperation, free exchange of business and technical information, responsive win-win decision-making and supplier profit sharing.

- **Maintenance and Reliability Engineering Group**—Because statistics indicate that up to 70% of equipment failures are self-induced, a major responsibility of Maintenance Engineering involves discovery of the causes of all failures. Reliability Engineering is a major responsibility of a maintenance engineering group. Their responsibilities in this area also include evaluating Preventive Maintenance Action effectiveness, developing Predictive Maintenance (PdM) techniques/procedures, performing condition monitoring/equipment testing and employing engineering techniques to extend equipment life, including specifications for new/rebuilt equipment, precision rebuild and installation, failed-part analysis, root-cause failure analysis, reliability engineering, rebuild certification/verification, age exploration and recurrence control. (see the following definitions for a description of some of these terms).

- **Root Cause Failure Analysis (RCFA)**—One of the most important functions of the Maintenance Engineering group is RCFA. Failures are seldom planned for and usually surprise both maintenance and production personnel and they nearly always result in lost production. Finding the underlying, or root cause of a failure provides you with a solvable problem removing the mystery of why equipment failed. Once the root cause is identified, a fix can be developed and implemented. There are many methods available for performing RCFA such as the Ishikawa, or Fishbone, diagramming technique, the Events and Causal Factor Analysis, Change Analysis, Barrier Analysis, Management Oversight and Risk Tree (MORT) approach, Human Performance Evaluation and the Kepner-Tregoe Problem Solving and Decision Making Process are some of the more common.

- **Failed Part Analysis**—Examination, testing and/or analysis by Maintenance Engineering on failed parts and components, removed from equipment, to determine whether the parts were defective or an external influence such as operating conditions, faulty installation technique or other influence caused the failure. Physical examination is often required in order to determine where to begin RCFA. For example when a bearing fails it must be determined, by examination of the bearing, the mode of failure. If electrical erosion/pitting is found, then stray ground currents (the cause of electrical pitting in bearings) must be found and eliminated.

- **Procedure Effectiveness Analysis**—The responsibilities of Maintenance Engineering for the establishment and execution of maintenance optimization involves the use of CMMS generated Unscheduled and Emergency reports and Planned/Preventive Maintenance reports to determine high-cost areas, establish methodologies for CMMS trending and analysis of all maintenance data to make recommendations for changes to Preventive Maintenance frequencies, Corrective Maintenance criteria and overhaul criteria/ frequency. They must also identify the need for the addition or deletion of PMs, establish assessment processes to fine-tune the program and establish performance standards for each piece of equipment. The Maintenance Engineering group also establishes adjustment, test and inspection frequencies based on equipment operating (history) experience. Additional responsibilities include the optimization of test and inspection methods and introduction of effective advanced test and inspection methods. Maintenance Engineering performs periodic reviews of equipment on the CM/PdM program to delete that equipment no longer requiring CM/PdM, or add to the CM/PdM program, any equipment or other items as appropriate. The Maintenance Engineering group also communicates problems and possible solutions to involved personnel and controls the direction and cost of the CM/PdM program.

- **(PdM) Analysis**—A major role of Maintenance Engineering is optimizing maintenance. One of the most widely used tools in this regard is Predictive Maintenance (PdM) to forecast necessary maintenance actions. Depending on the quantity and kinds of production equipment in your plant, the array of PdM techniques can range from as few as two or three to as many as ten or even more. Whether a PdM technique is outsourced or performed in-house, the results and recommendations must be analyzed by Maintenance Engineering and maintenance actions scheduled prior to predicted failure or out-of-specification condition.

- **Trending and Analysis of Condition Monitoring**—Condition Monitoring, actually a subset of Predictive Maintenance, usually involves the use of installed metrology (gages, meters, etc.) to derive the equipment's operating condition. Examples can be as simple as a differential pressure gage across a filter or the head-flow characteristics of a pump. Maintenance Engineering must establish operating limits for the condition(s) being monitored and trend the observed data, obtained from a log sheet or planned maintenance procedure, to determine when the operating limits will be exceeded so that required maintenance can be performed. This is referred to as condition based maintenance and can be both more effective and less costly than periodic or fixed frequency maintenance.

The foregoing provides a good, basic definition of Lean Maintenance by describing the activities and job responsibilities of those involved in the Lean Maintenance operation. Lean Maintenance is also about fundamental changes in attitudes and leadership roles. In the Lean environment the shop floor level employee is recognized as the company's most valuable asset. Management and supervisory roles change from that of directing and controlling, to a role of support.

The Lean Maintenance organization is a flat organization with fewer layers of middle management and supervision because, with the estab-

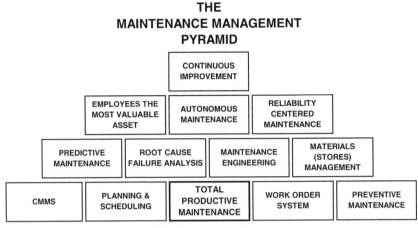

Figure C-15

lishment of empowered action teams much of their direction comes from within. The remaining supervisors spend the majority of their time on the shop floor providing technical advice and guidance and identifying first-hand the problems and needs of action teams.

The foundation elements, in particular TPM, must be in place before you can effectively build on the Maintenance Management Pyramid (See Figure C-15) with elements such as Autonomous Maintenance and before you can sustain continuous improvement.

Clearly a company transitioning to Lean Manufacturing will not have a sound basis of maintenance support without first implementing many of these necessary and fundamental changes in the maintenance operation. As the foundation of Lean Maintenance, Total Productive Maintenance (TPM) must be operating and effective, as shown by your Key Performance Indicators, prior to launching your plant's Lean Manufacturing initiative.

Written by Ricky Smith, Executive Director of Maintenance Solutions, Life Cycle Engineering, Inc., email: *rsmith@LCE.com*, 843-744-7110 ext. 350

Glossary

Definitions of Words, Terms and Acronyms Used in this Text

Note: (J) indicates that the term is Japanese.

Andon (J)	A system of flashing lights used to indicate production status in one or more work centers.
Availability	(1) Informally, the time a machine or system is available for use.

$$\{Availability = MTBF \div (MTBF + MTTR)\} \qquad (2)$$

	From the Overall Equipment Effectiveness calculation, the actual run time of a machine or system divided by the scheduled run time. Note that Availability differs slightly from Asset Utilization (Uptime) in that scheduled run time varies between facilities and is changed by factors such as scheduled maintenance actions, logistics, or administrative delays.
Critical Failure	A failure involving a loss of function or secondary damage that could have a direct adverse effect on operating safety, on mission or have significant economic impact.
Critical Failure Mode	A failure mode that has significant mission, safety or maintenance effects that warrant the selection of maintenance tasks to prevent the critical failure mode from occurring.
Current state map	Process map of existing practices. A visual method of succinctly recording the key aspects of the current structure or process in the whole or any part.

Failure	A cessation of proper function or performance; the inability to meet a standard; nonperformance of what is requested or expected.
Failure Effect	The consequences of failure.
Failure Mode	The manner of failure. For example, the motor stops is the failure—the reason the motor failed was the motor bearing seized which is the failure mode.
Failure Modes and Effects Analysis (FMEA)	Analysis used to determine what parts fail, why they usually fail, and what effect their failure has on the systems in total.
Failure Rate (FR) or (λ)	The mean number of failures in a given time. Often "assumed" to be $\lambda = (MTBF)^{-1}$.
Five Ss	Five activities for improving the work place environment: 1. Seiketsu (J)—Sort (remove unnecessary items) 2. Seiri (J)—Straighten (organize) 3. Seiso (J)—Scrub (clean everything) 4. Seiton (J)—Standardize (standard routine to sort, straighten and scrub) 5. Shitsuke (J)—Spread (expand the process to other areas)
Future state map	Value stream map of an improved process (non-value adding activities removed or minimized)
Jidoka (J)	Quality at the source. Autonomation—a contraction of "autonomous automation." The concept of adding an element of human judgment to automated equipment so that the equipment becomes capable of discriminating against unacceptable quality.
Jishu kanri (J)	Self-management or voluntary participation
JIT	Just-in-Time. Receiving parts, material or product precisely at the time it is needed. Avoids inventory pile-up.
Kaizen (J)	The philosophy of continual improvement, that every process can and should be continually evaluated and improved in terms of time required, resources used, resultant quality, and other aspects relevant to the process.
Kaizen Event	Often referred to as Kaizen Blitz–A fast turn-around (one week or less) application of Kaizen "improvement" tools to realize quick results.
Kanban (J)	A card, sheet or other visual device to signal readi- -ness to previous process. (Related—visual cues:

operating and maintenance visual aids for quick recognition of normal operating ranges on gauges, lubrication points, lubricant and amount, etc.)

Karoshi (J)   Death from overwork.

Lean Enterprise   Any enterprise subscribing to the reduction of waste in all business processes.

Lean Manufacturing   The philosophy of continually reducing waste in all areas and in all forms; an English phrase coined to summarize Japanese manufacturing techniques (specifically, the Toyota Production System).

Mean Time Between Failures (MTBF)   The mean time between failures that are repaired and returned to use.

Mean Time To Failure (MTTF)   The mean time between failures that are not repaired. (Applicable to non-repairable items, e.g., light bulbs, transistors, etc.)

Mean Time To Repair (MTTR)   The mean time taken to repair failures of a repairable item.

Muda (J)   Waste. ⇒ There are seven deadly wastes:
1. Overproduction—Excess production and early production
2. Waiting—Delays—Poor balance of work
3. Transportation—Long moves, redistributing, pick-up/put-down
4. Processing—Poor process design
5. Inventory—Too much material, excess storage space required
6. Motion—Walking to get parts, tools, etc.; lost motion due to poor equipment access
7. Defects—Part defects, shelf life expiration, process errors, etc.

Mura (J)   Inconsistencies (J).

Muri (J)   Unreasonableness (J).

Non-value adding   Those activities within a company that do not directly contribute to satisfying end consumers' requirements. Useful to think of these as activities which consumers would not be happy to pay for.

Overall equipment effectiveness (OEE)   A composite measure of the ability of a machine or process to carry out value-adding activity. OEE = % time machine available × % of maximum output achieved × % perfect output.

P—F Interval	The amount of time (or the number of stress cycles) that elapse between the point where a potential failure (P) occurs and the point where it deteriorates into a functional failure (F). Used in determining application frequency of Predictive Maintenance (PdM) Technologies.
Pareto analysis	Sometimes referred to as the "80:20 rule." The tendency in many business situations for a small number of factors to account for a large proportion of events.
PDCA or PDSA	Shewhart Cycle: Plan-Do-Check (or Study)-Act. A process for planning, executing, evaluating and implementing improvements.
Poka Yoke (J)	A mistake-proofing device or procedure to prevent a defect during order intake, manufacturing, process or maintenance process.
Predictive Maintenance (PdM)	The use of advanced technology to assess machinery condition. The PdM data obtained allows for planning and scheduling preventive maintenance or repairs in advance of failure.
Preventive Maintenance	Time- or cycle-based actions performed to prevent failure, monitor condition or inspect for failure.
Proactive Maintenance	The collection of efforts to identify, monitor and control future failure with an emphasis on the understanding and elimination of the cause of failure. Proactive maintenance activities include the development of design specifications to incorporated maintenance lessons learned and to ensure future maintainability and supportability, the development of repair specifications to eliminate underlining causes of failure, and performing root cause failure analysis to understand why in-service systems failed.
Process Mapping	Technique for indicating flows or steps in a process using standard symbols. Used to facilitate process improvements.
Pull system	A manufacturing planning system based on communication of actual real-time needs from downstream operations ultimately final assembly or the equivalent—as opposed to a push system which schedules upstream operations according to theoretical downstream results based on a plan which may not be current.

Reliability	The dependability constituent or dependability characteristic of design. From MIL-STC-721C: Reliability—(1) The duration or probability of failure-free performance under stated conditions. (2) The probability that an item can perform its intended function for a specified interval under stated conditions.
Reliability-Centered Maintenance (RCM)	The process that is used to determine the most effective approach to maintenance. It involves identifying actions that, when taken, will reduce the probability of failure and which are the most cost effective. It seeks the optimal mix of Condtion-Based Actions, other Time- or Cycle-Based actions, or Run-to-Failure approach.
Return on Investment (ROI)	A measure of the cost benefits derived from an investment. ROI (in %) = [(total benefits—total costs) ÷ total costs] × 100
Saki (J)	A rice wine, preferably served warmed
Sensei (J)	One who provides information; a teacher, instructor
TAKT Time	Cycle Time (G)—In Production it is the daily production number required to meet orders in hand divided into the number of working hours in the day.
Total Productive Maintenance (TPM)	A manufacturing-led initiative for optimizing the effectiveness of manufacturing equipment. TPM is team-based productive maintenance and involves every level and function in the organization, from top executives to the shop floor. The goal of TPM is "profitable PM." This requires you to not only prevent breakdowns and defects, but to do so in ways that are efficient and economical.
Value adding	Those activities within a company that directly contribute to satisfying end consumers, or those activities consumers would be happy to pay for.
Value Stream	The specific value adding activities within a process.
Value Stream Mapping	Process mapping of current state, adding value or removing waste to create future state map or ideal value stream for the process.
Vibration Analysis	The dominant technique used in predictive maintenance. Uses noise or vibration created by mechanical equipment to determine the equipment's actual condition. Uses transducers to translate a vibration amplitude and frequency into

electronic signals. When measurements of both amplitude and frequency are available, diagnostic methods can be used to determine both the magnitude of a problem and its probable cause. Vibration techniques most often used include broadband trending (looks at the overall machine condition), narrowband trending (looks at the condition of a specific component) and signature analysis (visual comparison of current versus normal condition). Vibration analysis most often reveals problems in machines involving mechanical imbalance, electrical imbalance, misalignment, looseness and degenerative problems.

Wiebull Distribution   A statistical representation of the probability distribution of random failures.

Index

motion, 110–111
overproduction, 108–109
processing, 110
reduction, 11
transportation, 109–110
waiting, 109
Watt, James, 1
Wiebull Distribution, 270
WO (Work Order) Discipline, 253
Womack, James P., 10
Work execution, 60–63
Work flow, 64, 66
best maintenance practices,
226–227
Work flows, defining, 11
Work order system, 227, 259
Work orders, 64, 66
assigning, 128
breakdown structure, 25
closing, 129
CMMS generated, 128

completing, 128–129
informally defining elements, 128
maintenance engineering, 128
priorities, 96, 98–99
production generated, 128
scheduling, 128
self-induced equipment failures,
233–234
tracking performance, 129
Work process
productivity KPIs (Key
Performance Indicators), 40
standardization, 128–129
Work sequence, 111
Work task, 212
Work type organization, 63
Working environment, 46
World class logistics support, 155

X
Xerox, 44